This is an introductory survey of the philosophy of science suitable for beginners and nonspecialists. Its point of departure is the question, Why should we believe what science tells us about the world?

In this attempt to justify the claims of science the book treats such topics as observation data, confirmation of theories, and the explanation of phenomena. The writing is clear and concrete with detailed examples drawn from contemporary science: solar neutrinos, the gravitational bending of light, and the creation–evolution debate, for example. What emerges is a view of science in which observation relies on theory to give it meaning and credibility, whereas theory relies on observation for its motivation and validation. It is shown that this reciprocal support is not circular since the theory used to support a particular observation is independent of the theory for which the observation serves as evidence.

Attractive features of the book include a glossary of technical terms and concepts commonly used in the philosophy of science, and a helpful guide to further reading.

READING THE
BOOK OF NATURE

Reading the
Book of Nature

An Introduction to the Philosophy of Science

PETER KOSSO
Northern Arizona University

CAMBRIDGE
UNIVERSITY PRESS

Published by the Press Syndicate of the University of Cambridge
The Pitt Building, Trumpington Street, Cambridge CB2 1RP
40 West 20th Street, New York, NY 10011-4211, USA
10 Stamford Road, Oakleigh, Victoria 3166, Australia

© Cambridge University Press 1992

First published 1992

Printed in the United States of America

Library of Congress Cataloging-in-Publication Data

Kosso, Peter.

Reading the book of nature : an introduction to the philosophy of
science / by Peter Kosso.

p.   cm.

Includes index.

ISBN 0-521-41675-2(hard). – ISBN 0-521-42682-0 (pbk.)

1. Science – Philosophy.   I. Title.

Q175.K8648   1992

501 – dc20   91–34429
                                                    CIP

A catalog record for this book is available from the British Library.

ISBN 0-521-41675-2 hardback
ISBN 0-521-42682-0 paperback

CONTENTS

ACKNOWLEDGMENTS

I am grateful to my students, especially the undergraduates in my epistemology and philosophy of science classes at Northwestern, who, without knowing it, got me started on this project. By patiently hearing me out, responding, and letting me know that clarity was appreciated and was compatible with philosophical interest, they convinced me to straighten out my account of science and get it down on paper.

I am pleased to acknowledge the Department of Philosophy at Northwestern University for assistance in the formative stages of the manuscript and for providing just the right atmosphere for writing.

Various people have commented on various drafts of the text, and I appreciate the suggestions they have made. My thanks to Paul Teller, Tom Ryckman, Philip Kitcher, and the unnamed reviewers.

Thanks as well to Cindy who, at the beginning, when I wondered whether to do this or something else, said to do what made me happy. Good advice, and I took it.

And to my parents, who let me turn in any direction I chose, even to philosophy, this book's for you.

INTRODUCTION

S CIENCE enjoys a lot of respect these days, if not always for the social value of its results then at least for the rigor and precision of its methods. It is an honor to say of a study or an argument that it has been done scientifically. This must be because we think that, in general, science and scientific methods are effective for getting at the truth. If pressed to articulate these feelings we might start by saying that the careful, accountable, and methodical approach of science prevents, as much as possible, the influence of personal bias, superstition, sloppiness, whimsy, sloth, and other human weaknesses that obscure the truth. Thus, when we claim to have done something scientifically, we speak with the authority of a truth-generating enterprise.

There has to be more to it than just being careful (careful to do *what?*) and accountable (accountable by what standards?), and we can appropriately ask why science has its special status as a supplier of knowledge about the world. Thus the unifying question in the pages that follow is, Why should we believe what science says about the world?

This is a question of justification of scientific knowledge. It asks not so much about *what* science claims about the world but more about *how* science proves what it does claim and why it gives us reason to believe that these claims are true. To ask about the justification for knowledge is not an attack on its credibility. There is no implication of skepticism here, and the request for proof is not a veiled suggestion that there is none to give. Instead, the concern for justification represents a requirement of responsibility that is the burden of any purveyor of knowledge. Not just any belief or

1

statement counts as knowledge, only those that can be justified as likely to be true.

Perhaps, in the case of science, finding the justification is as easy as giving credit for success. Science gains credibility from its phenomenal success. Airplanes fly. Smallpox has been done away with. Humble uranium gets turned into electricity, which in turn keeps the Bud Light cold. This record of success may be enough to give us responsible warrant to believe that science, by and large, gets at the truth about the world, but there are three reasons why this appeal to success is just a suggestive beginning to the analysis of scientific justification.

For one thing, the most interesting science – that is, the sorts of claims whose justification we are most interested in – is about things that nonspecialists can never check for themselves. The currency of science is largely of unobservable entities such as electrons, black holes, DNA molecules, tectonic plates, and the like. We can *see* that airplanes fly, but how do we know that the electrons in an accelerator or in an atom do what the physicists say they do? What counts as success of a theory that is about things that can never be experienced? The criteria and measure of success in this domain are the business of experts, the scientists themselves, and in this sense science is a largely self-regulating business whose success in crucial areas is self-proclaimed.

Again, this is not intended as a threat to the integrity or credibility of science. It is, though, a warrant to investigate further just what it is about science and doing things scientifically that justifies belief in the results being true.

A second reason to be skeptical of the direct association of success as justification is the undeniable fact that the scientific description of the world changes over time. It was once respectable and successful science to describe the universe as being full of ether, and for a very long time the scientific account of the heavens placed the planets and stars in crystalline spheres centered at and revolving around the earth. We no longer take these once-successful theories to be true. Such a dramatic turnaround should be sufficiently humbling to allow us to ask whether quarks and quasars aren't just today's flash in the pan, which may appear as object lessons in the introduction to a book written years from now on justifying scientific knowledge. An arrogance of the present (yes, but *now* we've got it right) can be avoided by focusing on a general method by which to tell when we've got it right.

The third motivation for looking beyond success is this: Even if these worries about the evaluation of success (it's left up to internal review; it's too ephemeral) could be put aside, there is a deeper question of interest here. What fuels the success? To say that something is done scientifically is not to say simply that it is done with success. Success may be the result, but it is not the methodological structure. What then is this structure, and what about it gives us reason to think it produces truths?

Putting the question in this way motivates the plan of presentation in the following chapters. The first step in the analysis will be an accurate description of the scientific enterprise, focused on the activity of justification of theories rather than their discovery. A clear account of the process of science will then be the evidence for deciding the issue of justification. Once the picture is clear as to how science works, the next step is to ask whether this way of working is conducive to producing the truth. The motivating goal is to be able to evaluate the justification of scientific claims, but first we need to know just what it is we are evaluating.

The specimen under observation here is going to be science at its best. We will not be concerned with fraud in science, episodes such as lying about experimental results, fudging the numbers, and other willful violations of the code of good scientific practice. The topic is the code itself and an assessment of its truth-conduciveness when properly followed. This abstract level of analysis is analogous to a political scientist's study of a social policy to see if it facilitates a just distribution of social welfare. Of course, there will be cheaters, but before dealing with these problems, the analyst wants to know the worth of the policy when it is properly enforced. Can it deliver as it promises, whether it promises justice or truth, when it is working at its best?

All this talk of justification of scientific knowledge and the scientific method makes it sound as if we have license to generalize about all sciences following a single method at all times. But surely the methods differ from scientific discipline to discipline. From astronomy to zoology, the diverse varieties of sciences study very different kinds of objects and very likely require different ways of theorizing and testing. Even within each domain it is unlikely that there is a single road to the truth, appropriate for all circumstances.

Just as surely, though, all these endeavors have *some* aspect of method in common, and it is this shared methodological ground that is the object of study here. What is it that allows all of these disciplines to be described as being done scientifically and as presenting characteristically scientific evidence? It is at this level of generality that we operate here, and I intend to demonstrate that the common methodological factor in the sciences (when done at their best, don't forget) is important as a truth-indicative ingredient. In other words, working at the high level of generality does not force the discussion to rise above all that is interesting and into the realm of only trivial generalizations. Granted that many of the features common to all the sciences are unimportant to evaluation of justification, there are nonetheless some common features that are significant to justification. But there is no need for you to believe this now, because the demonstration of this point is exactly the burden of what is to follow.

To be blunt, then, I do plan to generalize across the sciences and to proceed as if there is, at some level, a unity of method. I plan to generalize this much and then some by suggesting that the methodological mark of good science is what we take to be

4

the character as well of good common sense. In other words, as in science so too in life on the street. A close look at science is as a view of our less glamorous reasoning processes under a microscope, in that science is a slowed-down, more open and accountable image of what we normally do in coming to know about the world around us. Good science is an image of good sense.

I do intend to pay attention to a distinction between natural and social sciences, and the discussion here will be limited (though not strictly) to the natural side of the divide. The book of nature is of things without minds or the ideas, intentions, and emotions that are characteristic of minds. The focus on the natural world is not meant to suggest, though, that studying different kinds of objects, nonmental versus mental, necessitates wholly different methods of theorizing and justification. In fact, there is good reason to believe that the methodological model of the natural sciences also fits rather neatly onto the social sciences. I hope to motivate this belief by applying the understanding of natural science to examples of archaeology, indicating that this bridge discipline, with clear affinities to both natural and social studies, is methodologically similar to the natural sciences.

Furthermore, the model that will be used throughout for understanding natural science is one that is most often used in describing social sciences. This is the model of reading a text. Just as science builds models of the world, describing gases as crowds of colliding point-particles, for example, this study of science builds a model of the process of science as being methodologically like reading and interpreting a book, the book of nature. The method by which science comes to understand the natural world is very similar to the method used to understand the meaning of a text written in an unfamiliar language. Understanding the mechanism of nature is like understanding the plot. It is important to realize that this analogy between science and reading is intended to apply only at a methodological level. Though the methods of understanding and justification may be similar between the two activities, this does not imply that their subject matter is similar as well. In particular, the methodological analogy is not meant to suggest that nature has an author.

The translator of an unfamiliar text comes to understand its meaning through an attention to context. Recurring patterns in the symbols on the page prompt speculation about the meanings of passages, and most importantly, these speculations are testable and amenable to revision. Under the assumption that the text makes sense, that is, that the passages are consistent and they cohere, at least in small sections, into a cogent message, testing a speculated translation of a symbol is done by applying it to other occurrences in other contexts to see if it still makes sense. If the passage comes out as nonsense, the hypothesis must be revised. Through this back-and-forth of suggestion and revision, an understanding of the plot develops and is used to facilitate finding the meaning of passages by giving them a large context into which they fit. Thus the understanding of the whole text guides the interpretation of the parts, and the whole message is itself composed of the meaning of the parts.

This is what happens in reading the book of nature. Instead of marks on a page there are the experiments and observations done by scientists, but like the unfamiliar marks on a page, the observations are meaningless without some prior understanding, if only at the stage of revisable speculation, of what's going on in the story. In the context of science, the observations aren't evidence of anything without a theoretical description of their relevance. For example, the data of a particle detector are interesting only if we know already that a click means that an electron has passed through. Theories, such as the one that describes the link between clicking meters and passing electrons, represent an understanding of general aspects of how the world works and will influence the interpretation of individual pieces of observational evidence. Thus the big picture guides the interpretation of the parts. And of course the theoretical understanding of the big picture is built from and tested against the individual, observational parts.

This reasoning may sound circular, both in the case of translating a text and in that of acquiring knowledge of the natural world, but it is not necessarily a harmful circularity. In both situations, translating and knowing about the world, the project is to describe what is going on behind the scenes. The achievement of science or of translation is to make expansive inferences about things other than but significantly related to what is apparent. It's not interesting simply to describe the appearances, to reproduce or analyze the marks on the page. We want to know the story. We want to know what the marks mean. In science we will never get a look at the answer key, in the form of an authoritative translation of the book or even a dictionary, so we must learn to do the best we can with the available information. The developing theories and theoretically influenced evidence are all we have to go on to draw an accurate description of the world. Of course, somehow this cozy relationship must be regulated to insure a factual reading of the book of nature. What then constrains this process in the direction of truth? That is exactly the right question to ask, and it is time to get on with answering it.

A bit of advice to the reader: The form of this book is true to the message of its content. Individual chapters will make sense and attain their full meaning within the context of the whole work. This is just a friendly warning that the contents are not piecemeal, and parts taken in isolation will convey less information than those parts taken as they participate in the whole. But reading this book will be easier than reading the book of nature, at least in that this one has a lexicon, a glossary of key terms at the end. Terms that are included in this glossary, when first mentioned in the text, will be printed in bold type.

1

THEORIES

> "When *I* use a word," Humpty Dumpty said in a rather
> scornful tone, "it means just what I choose it to mean —
> neither more nor less."
>   — Lewis Carroll, *Through the Looking-Glass*

HUMPTY Dumpty knew what he was talking about. We are
in control of the meanings of the words we use, if not as
individuals then at least as a group, but sometimes words acquire
an authority and a persuasion that exceed our understanding of
them. Through our profligate use, a word may become influential even when its meaning is unclear and poorly understood.
When the link between meaning and emotive influence is
strained like this, the authority of language has gotten away
from us.

The word "natural" is a clear example of this. It means roughly
that something is not human-made, not artificial, but influence
has grown through advertising to suggest that anything natural
must be good, wholesome, and healthy. The link between what
we mean by "natural" and the authority it has over our behavior
is poorly understood, at best. Why are things that are not human
fabrications better, across the board, than those that are? More to
the point of meaning and understanding, what does it mean to
describe a product like shampoo, breakfast cereal, or even rat
poison as "all-natural"? And why is it better for me, and the
rat too, presumably, if I use all-natural rat poison? The word
"chemical" is another case where authority is wielded in the

8

absence of understanding. Sinister in suggestion though unclear in meaning, it becomes a club of conviction, as we opt for chemical-free products.

A third example of a term whose use and influence are much more common and secure than is its understanding is "scientific." Treatment of this case will last the rest of the book. It is a persuasive word, and though poorly understood it surely indicates something good in what it describes. Responsible use of the concept and responsible citizenship in a society under the influence of science require clarity in both the meaning of the term and the link between the meaning and its authoritative use.

WHAT THEORIES AREN'T

Still heeding the advice of Humpty Dumpty, we begin our pursuit of clarity in the concept of science with a discussion of the use of a key scientific term, "theory."

Though unsure of just what it takes to be a **theory** or what it means to be theoretical, people inside and out of science often use the concepts in a vaguely pejorative way, as in "But that's just a theory!" This suggests a distancing from reality and a lack of proof, an absence of good reason to believe a claim that is *just* a theory. The implicit contrast is between facts, which are true of the world, and theories, which are, well, who knows? It may even be that there is good reason to disbelieve a claim that is merely theoretical. Presentations of the debate between supporters of the theory of evolution and advocates of creation science can exemplify this attitude. (Note the use of the appellation "science" by the latter camp, indicating that this is for them and, they assume, for their audience a thumbs-up word, indicative of good, solid work.) Many representatives of creation science open the attack by describing evolution as just a theory. Good science, the stuff we should be teaching in schools, must be factual science.

The irony is that the major accomplishments of science, the things that come to mind when we reflect on what good, successful, responsible science is, are all what we describe as theories. Think about it. What has science accomplished? It figured out, for one thing, the phenomenon of universal gravitation, that all massive objects, whether on the earth or in the distant heavens, are forcefully attracted to one another. This describes a variety of happenings in the universe, from the falling of unsupported things around us to the motions of planets and stars, to the expansion of the whole universe. In sum, we, as modern and informed commentators on science, must be referring to the general theory of relativity.

Science also has a practical impact in its application to the difficulties in coping with the world. It has explained the unseen mechanisms responsible for transmission of diseases. Who doesn't believe in the germ theory? And it describes the constituents and events in the microworld with sufficient detail and clarity to support the fairly reliable development of solid-state electronics and, of course, nuclear bombs. We must be talking now of the quantum theory. If you want to know about earthquakes, where, when, and why they occur, and you want a scientific answer, you consult the theory of plate tectonics. And so on.

When you stop talking about theories, you stop talking about science. So, if being scientific is a good thing as far as methodical approach to the truth about the world is concerned, and theories are the primary product of the scientific enterprise, it cannot be that theories are, because they are theories, unlikely to be true or in want of whatever it is that science provides as proof. Being theoretical is nothing to be ashamed about.

Is it something to be proud of? If "theory" is not a term of censure, is it a term of praise, indicative of a credible and comprehensive description of some feature of the world? No, at least not if we pay attention to how the concept is used in science. Consider two common references to the concept of theory, caloric theory and kinetic theory. Both are descriptions of the phenomenon of heat, though they describe the underlying mechanism in very different ways. The caloric theory has it that heat is a substance (called caloric), a subtle fluid that seeps into and out of things making them hotter and colder, respectively. Being hot is like being wet; only the fluid is different, being caloric instead of water. The kinetic theory, in contrast, claims that heat is not a special kind of stuff but a kind of activity in the molecules of whatever stuff is hot. Nothing flows into a potato to make it hot. Instead, the molecules that make up the potato simply vibrate faster. There is no exchange of materials and, in fact, no such thing as caloric. Clearly, these two theories are in competition as the acceptable description of heat. One is more likely to be true than the other, but neither is more theoretical than the other. The likelihood of each being true is not assessed on the basis of how theoretical each is. The scales are not even coincident. To put it bluntly, it makes as much sense to talk about true theories as it does about false theories. Being theoretical is irrelevant to being true.

It is irrelevant as well to the distinction between being well proven and being new and speculative. The special theory of relativity, with its account of motion, mass, energy, and their interactions in space and time, has passed many scientific tests with excellent scores. It's a well-established theory. Contrast it with superstring theory, a relatively new suggestion about the basic structure of space, time, and everything therein. It is a provocative, though untested, description of the world, and it is a theory. The point is not that these two theories compete (they don't) but that they both fit comfortably within the concept of being a theory even though they differ significantly in their status of being tested.

Perhaps, if "theory" is not laudatory in the sense of being likely to be true, it is a credit to the structure of a claim in that the account is a comprehensive, complete description of some aspect of the world. It may take several interrelated concepts to cinch up all the loose ends in describing an electron, say, or a tectonic plate, and only the whole package of claims, models, even experimental procedures, counts as a scientific theory. We do sometimes regard theories as complex, multifaceted things, but it is hard to say exactly what must be included to be a complete theory and thus to be a proper unit of scientific evaluation in this sense. It is also unnecessary. Since the project is to assess the justification of scientific claims, it is best to keep a mind open to the possibility of parts of these multifaceted things, that is, individual claims, being justified or not. This advocates an interest in concepts of theories as individual, even isolated, claims and allowance for language such as "the theory that cosmic rays originate in nuclear reactions in outer space."

This means that you can have theories that are general or specific, comprehensive or selective. Indeed, normal use of the language reflects this openness as we speak of the theory of nuclear fusion as well as a more comprehensive theory of the nucleus. You can have a general theory of the emergence of civilization or a theory of the significance of Stonehenge. You can have a theory that the dinosaurs were done in by catastrophic climatic changes brought on by a huge meteorite. You can even have a theory that it's the guy next door who keeps stealing your Sunday newspaper.

To summarize, none of these features – likelihood of being true, generality, or comprehensiveness – constrains the concept of what it is to be a theory. This generous latitude in the use of the concept devalues its authority. To say of something that it's a theory is neither a credit nor a discredit. There will be good and bad theories in terms of their credibility, but they will be theories nonetheless. What is needed is a way to decide whether a particular claim warrants belief, but simply pointing out that "it's a theory" won't help. We need further information as to whether it's a *good* theory.

12

All of this has been about how not to limit or overemphasize the importance of the concept of a theory. It has been about what "theory" doesn't mean, but surely the concept has some limits. What *does* "theory" mean?

WHAT THEORIES ARE

In answering this question it is best to pay attention to the earlier comment that described theories as the primary products of the scientific enterprise. What then is the goal of science? What is it in the business of producing? But most generally, science aims to provide an account of what's going on behind the phenomena we experience. By describing objects and events that are not apparent, it tries to make sense of the happenings in the world that are apparent. Science aims to understand the composition and underlying causes of the manifest phenomena and thereby make sense of the phenomenal world we inhabit. This pursuit of understanding could be motivated simply by an urge to satisfy curiosity, or to explain the world, or by a more pragmatic desire to deal with the natural and social environment. Any effort to control the events in the world requires an ability to predict and anticipate happenings and to prevent the undesirable or bring about the desirable. These skills demand an understanding of causes, the unseen mechanisms that determine the events we experience. Thus, whether it is pure or applied, if it's science it must seek an understanding of the underlying constitution and causes of phenomena.

Science is not after simply a careful, methodical account of the observable world. It is not meant to be but a refined and accurate description of phenomena, the best possible picture of the apparent objects with a precise schedule of their comings and goings. Certainly this sort of detailed and accurate portrayal of the manifest world is important as a necessary means for discovery and testing in science. Discovery and testing of what? Theories. The accurate observations serve in support of the account science gives of what the world is like in the realm that is too small, too distant, or too far in the past to be experienced.

If science is applied, for example, to the phenomena of acid rain and the dying of forests and lakes, the result is more than a detailed list of the effects. We want to know just what's going on here, and that is exactly what we get. The scientific report is both useful and interesting insofar as it describes the unseen constituents of the problem, the acids and their chemical components, and the basic events of their production and coming together in the atmosphere. This account of what is happening beyond what is directly apparent is the product of science, and it is why we turn to science for help and for understanding.

Earthquakes are another example in which the role of science is to describe what cannot be directly experienced. We can all feel an earthquake, and what we expect of geology is to explain why there are earthquakes and explain it in terms of what it is about the earth that we cannot feel or see that causes the damn things.

The most basic science-classroom demonstrations, those intended to teach what science is all about, are the same sort of thing. My favorite involves rubbing a balloon on someone's hair and then touching the balloon to the wall. It sticks. We start doing science when we ask, What's going on here? But this is not asking for a description of the observable phenomenon. This is asking for an explanation of what makes it happen. What is it about the basic composition of the balloon, the hair, and the wall, and what is it about the event of rubbing that causes the balloon to stick? Maybe a thin skin of old, dry rubber is scraped off when the balloon is rubbed on the hair, exposing the fresh, sticky rubber underneath, and that's what makes it adhere to the wall. Or maybe the important action is at an even smaller scale, and the rubbing actually scrapes microparticles of electric charge from the hair, making the balloon itself electrically charged and attractive to charged particles in the wall. It is this sort of description of the unperceived factors that are responsible for the phenomena, this kind of behind-the-scenes account of the world, which is the accomplishment of science.

These are theories. A theory is an answer to a "what's going on here?" question. It is an account of the underlying composition and the unseen causes of the world as we experience it. It is a very liberal concept in that any description of the unexperienced world that is part of what accounts for and helps us understand the experienced world is a theory. There are general theories and very specific theories. Some are well proven, others fairly well disproven, and others still are unproven because they are untested. They have in common the feature of speaking about things beyond the evidence. They take risks in telling us more than we can see for ourselves, and it is exactly this feature that makes them both interesting and in need of justification.

This is not to suggest that there is a clear distinction between theories and facts. It would be misleading to study science by looking for a sharp fact–theory dichotomy, and it would lead to a distorted view of theories, facts, and science in general. This will be made more convincing in Chapter 6, in which the process of scientific observation will be discussed. For now it is enough to be warned that in no aspect of science will the certainty that is characteristic of facts be available. In other words, if indubitability is what it takes to pass from being a theory to being a fact, then there are no graduates and there are only theories. The substantive claims in science are those that speak beyond the obvious, and this brings on a degree of uncertainty. The interesting results are the risky results, and there is no space between trouble and trivia.

The risk is to be minimized, of course, and this is exactly the burden of justification. Theories come in degrees of credibility, and individual theories change over time as they become more or less verified, but the process of justification does not make them more or less theoretical.

It is sometimes helpful to group relevant and cooperating claims into a set. This is what is going on when we talk about, for example, the quantum theory or the big bang theory, multifaceted descriptions of particular features of the world. But responsibility requires that each claim or each model stand the test of justification, and it is unnecessary to focus on whole theories (in this sense of complete accounts) to see if they pass the test wholesale. For this reason, regard theories, insofar as they are the primary subjects of justification, as individual claims.

It will be helpful to point out that there are two kinds of theories within this conception. That is, in understanding the structure of science and its methods of justification, it is useful to distinguish between two kinds of claims about unobservable objects and events. (The use of this term "unobservable" will have to be clarified and somewhat regulated. Between now and Chapter 6 feel free to let common sense be a guide as to what is observable and what is not.)

One sort of theory will be of claims that are only about unobservables. These will be descriptions of the occupants and activities of the unobservable world with no mention of their impact on the observable world. For example, many of the claims made in the physicists' account of elementary particles make no contact with the world of experience. All baryons (protons, for example) are made of quarks. That describes two kinds of unobservable entities and their relationship of composition. Still more is learned of the affairs of particles from the claim that forces of interaction between them, as occur during collisions, for example, are caused by the exchange of still others, so-called virtual particles. This is a provocative and maybe even accurate account of the microworld, but at this stage it informs of a world isolated from that of experience. The same is true of descriptions of the distant past, reconstructions of the appearance of buildings and pottery, and of political and economic behavior. All claims of this sort describe solely what the describers cannot experience.

16

The other type of theoretical claim is needed to break this isolation. This requires claims that are about the interactions between unobservables and observables. In this category are accounts of the events in the past that have influenced the features of the traces found in the present, whether as surviving texts, archaeological remains, or evolved social practices. Other examples include descriptions of the causal impact of the microworld on the macroworld (and vice versa). These sorts of statements bridge the gap between what we experience, the manifest image of the world, and what we cannot experience but are interested in, the scientific image. And clearly, theoretical claims of this variety will be of essential importance in the understanding of scientific evidence and the process of testing.

Examples of such claims are easy to spot in the experimental activities of scientists. A charged particle, like a proton, will ionize some atoms as it flies by and, in a supersaturated vapor as created in a cloud chamber, visible bubbles will form at the ions. In this way the microparticle leaves its mark on the world of experience and we can keep track of aspects of its behavior. Black holes, since they gravitationally attract light, will bend light rays to the same effect as a lens, causing a visibly altered image of stars and galaxies near their position in the sky. What we cannot see has a noticeable influence on what we can. A ceramic pot, used, broken, and discarded in archaic Greece, gets inadvertently mixed with manure in the animal pens, and pieces are spread with the manure in the fields, only to be recovered and examined by a modern archaeologist who has a theory that the sherd came not from a pot used here in the field, but from one used back at the farmhouse.

These are the imaging theories. Just as the theoretical account of a microscope attests that what is seen is reliably informative of the specimen, these theories of the second kind assure that what we experience, the streak in a cloud chamber or the potsherd, is an informative image of its causal predecessor, the particle or the pot. It is a distorted image in the sense that protons don't look like vapor trails, and, of course some information about the proton does not show up in the image, but the distortions are taken into consideration. As long as we know that they are there, and it is the business of these theories to point them out, they can be taken into account and they will not deceive.

Our situation with respect to the unobservable world of scientific images has us, as perceivers, at one end of a causal chain. At the other end is the object of interest. The archaic pot, through a sequence of causal interactions including cultural reuse, weathering, and erosion, produces a sherd in a field, and all the archaeologist sees of it is the last step, the sherd. A proton causes ionization, causes bubbles, causes a vapor track, and all the physicist sees is the track. If the sherd or the track are to be evidence of anything, there must be a theory to describe the other end of the causal chain as well as all the intermediate links. Thus both kinds of theories, about unobservables only (such as how archaic Greeks manured their archaic fields, or how protons ionize atoms) and about the relation between unobservables and observables (pots used for cooking are of coarse fabric but with thin walls, and ions initiate the growth of visible bubbles), are used to reconstruct the interactive chain from object of scientific interest to manifest image.

What these claims have in common, what makes them all theoretical, is their reference to what is unobservable. They describe more than what is apparent, and in doing so they draw the question of justification. When is a theory to be believed? Or better, to what degree is it to be believed?

HYPOTHESES AND LAWS

The factors discussed earlier, the generality, comprehensiveness, and degree of verification of a claim, were dismissed at the time not because they are unimportant but because they were not relevant to the determination of what it is to be a theory. They are nonetheless crucial to an understanding of the process of science, and they coincide with other key concepts used in describing the scientific method.

As pointed out before, some scientific theories have been exhaustively tested and others have not. The former have gone through whatever it is that is the answer to our question of justification. Those that have not or that are just in preliminary phases of testing are hypotheses. To say of something that it is a **hypothesis** is to report on its justification status by describing it as untested or only minimally tested. It is a disclaimer, which says the claim is just suggestive and is asserted without strong commitment. There is nothing irresponsible in suggesting accounts of the world that, as yet, lack justification, provided this status is clearly marked. This is the usefulness of the term "hypothesis."

Clearly this feature of scientific claims comes in degrees. We will soon find that justification accumulates without ever reaching the stage of certainty. For this reason it is preferable to apply the adjective "hypothetical" rather than the noun "hypothesis" to the description of science and to understand not a dichotomy between what is and is not a hypothesis but the measure of a claim being more or less hypothetical. Science, as a publicly accountable enterprise, requires that its products bear the label of "hypothetical" as a warning if they are tentative and untested. This is a label that wears off slowly, rather than being peeled off at once. The hypothetical status of a theory changes, indicating that this status is not an intrinsic feature, descriptive of what the theory is about or what it says or the form in which it says it. Being hypothetical is a reflection of the theory's historical relation to the activities (or inactivities) of the scientific community. You can't tell merely by looking at a theoretical claim just how hypothetical it is. You must be told explicitly.

Sometimes the term "hypothesis" is used in a case-specific way to identify, in a particular experiment, the principal claims being tested. This is to distinguish the subject of testing from the other theories used to account, for example, for the principles of the experimental apparatus and proper conditions. If the cloud chamber is used to investigate the behavior of protons in a magnetic field, the hypothesis is something like this: Protons follow a right-handed spiral trajectory when moving through a magnetic field. There are other theories, as discussed a few pages back, at work in this experiment as well, but they are not the ones the experimenter intends to test. Even with this use of the concept of hypothesis, to be hypothetical is to be in need of testing.

There is another important variation among theories, and that is their degree of generality. Some theories are more general and hence more interesting than others. Theories can be general insofar as they ignore such factors as spatial location (apply anywhere), time (hold anytime), or other features that are irrelevant to the process described. They do this by specifying which factors *are* relevant to the particular behavior. For example, it is true of all unsupported objects very near the surface of the earth that they speed up as they fall. Their speed increases at a rate of 9.8 meters per second for each second they are in free fall. This applies regardless of the color of the object, how it was made, where you bought it, the time of day it is dropped, and so on. These features are irrelevant. What *is* relevant is that the object has mass (though how much mass is not a factor), that it is unsupported, and that it is near the surface of the earth. The theory generalizes over all objects with these properties.

Better yet, or at least more general, is the claim that all massive objects, wherever they are, attract each other with a force that is proportional to the product of their masses and inversely proportional to the square of their distance of separation. This covers a larger class of objects and events, and it specifies fewer relevant properties required for membership in the class. Its contribution

20

to our understanding of the world is that it identifies the property of things (they must have some mass) relevant to the phenomenon (their attractive force and its resulting motion). It identifies the *kind* of thing relevant to the phenomenon. These theories, which generalize and describe kinds of things, are **laws.** We speak of the law of free fall, itself a chapter in the law of gravity. There are plenty of other examples. Just one more is Coulomb's law, which claims that any electrically charged object exerts a force on any other electrically charged object. The relevant kind of thing here is anything that is electrically charged, and it is linked to the phenomenon of force.

Unlike the feature of being hypothetical, we should be able to tell by looking at the presentation of a theory whether or not it is lawlike. It must be in the form (perhaps implicitly) of a generalization. All *A*'s are *B*. The use of the term "law" here is in one sense much like its use in society. When we say it's the law, we mean it applies to everyone. There are, of course, many important dissimilarities between laws of nature and laws of society. The latter are the convention of (a few or many) people and can be broken or altered at will. Laws of nature, by contrast, are intended to reflect the ways of the world, which are discovered rather than imposed by us and which are largely unbreakable.

We *should* be able to recognize lawlike statements by their generality, but it's not so easy. How can we distinguish real laws from merely accidental associations? Consider the following examples. Pure copper conducts electricity. That's a law of nature. It is an implicit generalization about all samples of copper, and it associates a kind of thing (copper) with a particular behavior (conducting electricity). But here is another generalization that is *not* a law. No samples of pure copper exceed 100 million kilograms in mass. This is general in that it describes all samples of copper and claims they are under 100 million kilos. But this feature of mass is an accidental attribute of copper in that there is nothing, as far as we know, about the nature of copper that puts a limit on the

21

size of a sample. It turned out, in the chaotic mixing of elements that formed the universe, that no sample of copper that big was formed. But it was just an accident.

And consider the generalization that everyone in this room is over ten years old. That's not a law either. It identifies a particular kind of object (persons in the room), but it is not a *natural* kind. There are no features exclusive to this group that are important to the workings of the world, and there is no causal connection between being in the room and being over ten years old in the way that there is between being electrically charged and being forceful on other charges. Electric charge *causes* the force, but being in the room doesn't cause you to be old. The generalization about the ages of people in the room is most likely just an accidental association. It's not a law. If it were a law it would give me license to make predictions like "The next person to come into the room will be over ten years old." This prediction, however, is unwarranted since being over ten years old was just a circumstantial feature of the people now in the room. The law of conductivity of copper, on the other hand, does warrant the prediction that the next piece of copper will conduct electricity. Laws support predictions; accidental generalizations do not.

So the assessment of whether a certain claim is a law cannot be made simply by checking the form of the claim. It will depend as well on other things we know about the situation, that is, on other theories. It requires an informal preview of what is likely to be relevant to the phenomenon in question. In this way, atomic and solid-state physics indicate that atomic structure, that is, the identity of an element such as copper, is a likely factor in electric conductivity, but nothing links the nature of copper to a limitation on the mass of a sample. Similarly, nothing in my understanding of the room says that being over ten years old is relevant to occupancy. The determination of lawlikeness, as used in the previous examples and generally in our description

22

of science, is always sensitive to the theoretical context in which the decision is made.

At least the feature of being lawlike is relatively stable over time. This label is not supposed to wear off. Once a law, always a law, almost. In fact, what is assessed to be a law can change as the accepted theoretical description of the world changes. If background theories change, what may once have been assessed as an accidental association could now be understood as a relevant, lawlike connection. In the time of Aristotle, for example, to generalize about motions of things in the sky and on the earth was to group two distinct kinds of phenomena, and any similarities between the two would be viewed as accidental and irrelevant to the understanding of motion. There would have to be two sets of laws, one of celestial motion and another of terrestrial motion. Ever since Newton though, and the discovery that what is relevant to motion is not the terrestrial—celestial distinction but simply the property of mass, it makes perfect sense to generalize over all objects in motion and to regard such generalizations as describing laws of nature.

The lesson, of course, is to resort to an appeal to accident as an explanation of association only as a last resort. But what is important here is simply to appreciate that the concept of law in science is descriptive of the generality and scope of a claim but not its status regarding proof. A law is not an irreproachable, indubitable sort of claim, the deciding word in any dispute. Laws, like any other interesting claim in science, are answerable to the question of justification.

HYPOTHETICAL THEORIES, HYPOTHETICAL LAWS

The three concepts discussed so far, theory, hypothesis, and law, describe different aspects of scientific claims and are not exclusive of each other. They represent different features of claims, just as size, color, and shape represent different features of a

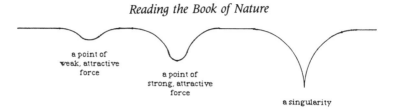

Figure 1.1. The strength of gravitational force is dependent on the curvature of the spacetime manifold. Here, only one of the four spacetime dimensions is shown. A singularity occurs where the curvature is so tight as to buckle the manifold.

physical object, and though there is no sense in describing something as both big and little or both square and round, it makes fine sense to describe it as big and round. Similarly, there is nothing silly or nonsensical in speaking of a hypothetical law or a hypothetical theory. This is no transgression of logic or language.

Here, in fact, is a good example of a hypothetical law, that is, a theory (a claim about events that are not apparent to experience), that is both lawlike (a generalization about a natural kind of object) and hypothetical (not well tested). It is cosmic censorship. The general theory of relativity describes the curvature of spacetime as being associated with gravitational forces. The sharper the curve, the stronger the force. At points called singularities, where the curvature is so sharp that the spacetime actually buckles, the force is so strong as to tear apart elementary particles and to perpetuate events inconsistent with our theories of the normal world (see Figure 1.1). Such a singularity can occur in nature, according to the theory, but it could be surrounded by an event-horizon, a one-way causal shield that prevents both our seeing these crimes against nature and their having any effect on the outside world. This is the situation of a black hole. It is possible as well for there to be singularities without event-horizons, naked singularities, possible at least

24

according to the general theory of relativity. So cosmologists have considered adding an additional claim to the current description of the universe. Simply, all singularities are in fact surrounded by event-horizons. This abhorrence of nakedness is the principle of cosmic censorship, but it is only at the stage of suggestion, and what is called for at this stage is not instant belief but a further suggestion as to how to test it. It is, in other words, a hypothetical, lawlike theory.

Conversely, there is nothing unreasonable in having a nonhypothetical law or nonhypothetical theory. Snell's law is a nice example. This is the account of the bending of light rays as they pass from one material into another, as from water into air, or air into glass. It specifies not only that the light bends but in which direction it bends and by how much. It all depends on the index of refraction, a property of the material through which the light is traveling. Here then is a theory (insofar as it describes the index of refraction) that is lawlike and very well tested. The best way to put it is that it is very low on the scale of being hypothetical.

The point of this discussion of theories, laws, and hypotheses is twofold. For one, these three concepts are orthogonal; they describe different features of the claims scientists make. In no sense does an idea graduate through the ranks of being hypothetical and then a theory and at last a law. Laws when conceived are hypothetical and are laws nonetheless (nor the more) as testing wears the hypothetical label fainter and fainter. And they are theories all the while.

The second point is that "theory," at least as it is used here, is a very broad term. It is neither laudatory nor pejorative. Since so many things count as being theories they are relatively easy to come by, easy, that is, to discover or create. The hard part is to then justify them, and although there may be lots of theories around, not all of them are good theories. Not all of them warrant our belief.

25

GOOD AND BAD THEORIES

By good or bad theories I simply mean here that there is or is not good reason to believe that the theory is true. Good theories are justified theories. This, of course, comes in all degrees, all degrees, that is, except perfection. No theory reaches the stage of certainty, of being beyond all doubt and immune from change. But the degrees of justification are exactly what we are interested in here, and we will get on with the project of understanding science when we get on with assessing the virtues theories can have.

So it is not simply the production of theories that makes science distinctive, that endows it with authority and is responsible for its success. Just coming up with theories is not what is scientific about science. Rather, it is the responsibility, declared and dispatched, to separate the good theories from the bad overtly, or at least to scale the better over the worse. It is this process of justification that is at the heart of what it means to be scientific and why that is something to boast about. The understanding of science should focus, then, less on exactly what it is to be a theory and more on the virtues characteristic of a good theory.

2

INTERNAL AND EXTERNAL
VIRTUES

THEORIES are like apples; there are good ones and there are bad ones. A good apple is flavorful and nutritious. A good theory is one that we have good reason to believe is true, or at least likely to be true. Apples have all sorts of features that are indicative of goodness and that can be used at the store for making smart choices. An apple's color, shininess, shape, and firmness are related to its flavor and food value. Similarly, theories have features that are indicative of their truth, and the task of justification is to identify these features and use them to guide choices as to which theories to believe. In this way, responsible theory choice and responsible science are not unlike responsible shopping. The big difference is that you get to go home and eat the apple and find out if you were right and it really does taste good. With theories, however, there is no feature that is the taste of truth, at least not that we get to sample. There are only features that indicate truth. You never get to go home and see unambiguously that atomic theory, for example, is true. But you do the best you can.

Theories have lots of different properties and can differ one from another in a variety of different ways. A few have been mentioned already, namely, the degree of generality and hypothetical status of a theoretical claim. An account of the properties of a theory is an answer to the inquiry, "So tell me what this theory is like." There are plenty of things to say, that is, plenty of features to mention. There is the feature of how it was discovered, by whom, at what time of day. Some theories have the feature of being proposed by Einstein; some lack this feature. Some have the feature of being liked by Einstein; some don't. There are also features of the content of a theory or of its form. Being logically consistent is a feature of form. Making reference to action at a distance, or to evil spirits, or to gluons are distinct features of content. There are even pragmatic features. Some theories are likely to be money-makers; others are not.

Clearly, some of these features of theories are irrelevant to an analysis of scientific knowledge and irrelevant to doing science. We want to focus on the important features, those that meet the following two criteria. First, they must be relevant to the likelihood of the theory being true. That is, they must be reliable indicators of truth so that our seeing whether the feature is present or not will be part of our warrant for believing the theory. Some theories have the property that you learned them on a Tuesday, but that is unimportant because it is neutral to their being true or false. Some theories have the property of describing goblin maintenance men to keep the world in good working order. That's an important feature because it helps us decide whether the theory is true or false.

The second criterion of importance is that the property is something we can in fact evaluate. The information must be accessible to us if it is to be of any use. To get back to apples for a moment, even if it's true that apples with darker seeds taste better, this isn't any help there in the store where information about the seeds isn't available. Similarly, a theory about the demise of dinosaurs is true if it is an accurate picture of the past events, but the feature of being an accurate picture is not something we can evaluate. We can't see what this theory is a picture of, and so we can't assess its accuracy. Much as we would like to use accuracy-of-picture or correspondence-to-facts as features that indicate truth, information about these features is unavailable. Features that are important to the justification of theories must be both truth-conducive and accessible.

28

The same standards are used in evaluating the testimony of a witness at a trial and in justifying a verdict. There are all sorts of features of a witness's testimony. It could be very loud or softly spoken, consistent or full of contradiction. Maybe it was given on a hot afternoon. Perhaps it is full of detail about the defendant. Some testimonies have the property of being fabricated in collusion with the defendant, and some, not necessarily others, have the property of being an accurate account of the crime. Note that some of these features of the testimony are what would be included in a reporter's description of the trial. Others, though, cannot be responsibly reported since their assessment relies on inaccessible information. We can't really tell whether the testimony has the property of being an accurate account of the crime, given that, in the context of the trial, the facts of the crime are not apparent. Responsible jurors must attend to those features of the testimony that can be evaluated with the available information, and of those, some will be helpful in determining the likelihood of whether the testimony is true or not. Start with the easy one, logical consistency. If the testimony contains contradictions, it can't be the whole truth and nothing but the truth. Then check for richness of detail. If the testimony is full of detail it is probably a report of the facts. Most people just aren't clever enough, or are too lazy, to fabricate their own details. The demeanor of the witness is an important feature too. A calm, confident witness who faces the jurors and the defendant during testimony is probably telling the truth, more likely, at least, than some shifty-eyed fidget who won't look you in the eye. These aspects of the testimony are more reliable indicators of its truth than are such features as being given on a hot afternoon, or on a Tuesday, or by a handsome witness or one who uses the word "really" altogether too often. These latter considerations are irrelevant.

The responsibility of the jury is to decide the truth of the criminal matter and to do this by finding and using the available, truth-conducive features of the testimony. The responsibility of the scientist is much the same, to decide the truth of the matter about the natural or social world. The responsibility is carried out in a similar way, by finding and using the available, truth-conducive features of the theories. So too is the task of those who study science.

INTERNAL AND EXTERNAL FEATURES

It will be helpful in studying the features of theories to divide them into two groups. This will be just a preliminary distinction, subject to future changes, but it will be useful for identifying and evaluating the important features. Later on we will be forced to question the strictness and clarity of the dichotomy, though a more or less continuous spectrum may endure. It is not unnatural to understand a spectrum (such as temperature) by first understanding the features of its two ends (hot and cold).

Initially, then, features of theories can be grouped into those that are **internal** and those that are **external.** Internal features are those that can be evaluated without having to observe the world. That is, they don't depend on observations to check correlations between what the theory says and what the world is like. In this sense they are very economical since there is no need for experiments or laboratories to check for internal features of a theory. They can be evaluated by studying books, that is, by looking at the structure of the theory and its conceptual relation to other theories. Logical consistency is a clear example of an internal feature. Some theories have it; some theories don't. Checking for logical consistency is not an experimental procedure. If my theory of tomorrow's weather, for example, is that it will both rain all day and not rain all day, you don't have to look outdoors to realize that something is wrong. It's inconsistent, and this is evident from the internal structure of the theory.

30

External features, on the other hand, are relevant to the theory's relation to the world, and they require observation for their evaluation. For example, some theories have the property of making true predictions; some do not. It is impossible to tell whether a theory makes true predictions without actually doing the experiments and observing the outcome. It's not essential that you do these things yourself, but someone in the community of knowers must have a look to see how the experiments go if anyone is to claim responsibly of a theory that it makes predictions that come out right. Thus external features report contact with the world and therefore demand observational evaluation.

The courtroom-testimony analogy will help to clarify this distinction between internal and external features. Internal features are those properties of the testimony one could detect just by listening to the testimony and to that of other witnesses without checking the correlation of the testimony to objects or events in the world. If the witness says, for example, that on the night of the crime it was dark and raining heavily but says later under cross-examination that on the night of the crime the moon was visibly full, the testimony has the feature of being inconsistent. No one has to have seen the weather that night to realize that this is inconsistent. As another example of an internal feature of the testimony, suppose one witness says that it was very dark at the scene of the crime. This report then explains the account of another witness who claims she didn't see anyone in the back seat of the car. These two testimonies have a feature of explanatory coherence. They make sense together, and this is apparent just from hearing them. Again, no observation of the car or the scene of the crime is necessary to see that the two testimonies cohere.

Testimonies will also have external features, and that is why physical evidence and visits to the scene of the crime are important parts of a trial. If the witness describes the criminal as being skinny and having a mustache, then there is only one way to tell if this testimony has the property of being an accurate description of the defendant, and that is to look at the defendant. Some member of the group responsible for deciding what's true, that is, some member of the jury, must perform the experiment and have a look.

The choice of terminology for this internal–external distinction may seem a bit mysterious as it is not obvious just what these features are supposed to be inside or outside. The internal features, since they do not speak of contact with the outside world are, in this sense, in the mind, at least the collective minds of the scientists or the jurors. They are discernible without the aid of the senses, our line of contact to the physical world. So, to evaluate the internal features of a theory is to think about it and see if it makes sense before you spend good time and money on the experiments to test it. To see if it makes sense, one compares the theory with other theories in mind, both in the individual's mind and in the collective mind of the group. External features, on the other hand, require information from the external world, from things other than ourselves. Evaluating these properties demands an influence from objects and events outside of our own minds and beliefs. Sometimes you even have to go literally outdoors and get dirty in order to check on external features. You never have to do this with internal features.

There are other pairs of terms that come close to describing this distinction, but none succeed quite as well as internal and external. We might consider, for example, speaking of theoretical virtues in distinction from experimental virtues. This is a close match to the internal–external distinction, but it is misleading in some important cases. Logical features, being free of contradictions, for example, do not fit neatly under the heading of being theoretical. They don't fit at all in the category of experimental. We could squeeze them in with the theoretical (we are, after all, in control of our own language), but then where do we squeeze the feature of making lots of predictions that will be amenable to observation? This is an internal feature because the fact that the theory makes these predictions can be discerned by analyzing its structure and its implications. Whether or not these predictions come out to be true is another story. The point is that the feature of making predictions has the sound of being experimental, yet it is internal, indicating that the theoretical–experimental distinction could precipitate confusion over internal and external features.

Another potential expression of the relevant distinction between kinds of features focuses on the difference between features of coherence and features of correspondence. Philosophers often use these concepts to describe features of beliefs we have, whether or not the beliefs deal with issues of science. **Coherence** is descriptive of a whole system of beliefs and indicates that the constituents fit together in some way. It describes relations between one belief and other beliefs. **Correspondence** describes the relation between one belief and the world. It's one thing for my belief that the man next door is stealing my Sunday newspaper to fit nicely with my other beliefs that he is a cheapskate and a crook; it's quite another for my belief to correspond to the fact of the matter as to whether or not he is stealing the newspapers. This distinction between coherence and correspondence is very similar to the separation of internal and external features, so close that we will at times drift into this terminology. The beliefs we have and that are analyzed by philosophers are just our theories about the everyday world, and questions about science are relevant to questions about knowledge in general. For now, though, we shall stick with the internal–external description. The idea of coherence implies an analysis of whole systems of theoretical claims to the exclusion of individual claims. We should leave open the possibility of internal features of individual theories.

The important, practical consideration in determining the value of any feature of a theory, whether it is internal or external, is whether it can be evaluated. We will want to use these features as a measurement of how good the theory is. They will be the hall-marks of acceptability and so are of no value if they are hidden or incomprehensible.

VIRTUES

Like the members of the jury, scientists need to focus on the fea-tures of a theory that indicate that it is a good theory. These are virtues. These are the important factors for comparisons of theo-ries, that is, for deciding between competing accounts of the world. There may not be an absolute scale of the goodness of a theory on which, for example, the heliocentric description of the solar system is a just-plain-good theory. But there ought to be a relative measure, a way of comparing theories, by which the heliocentric theory comes out as being better than the geo-centric account.

Different aims of theorizing will work with different standards of goodness. There are various ways of answering the question, Good for what? If the aim of science, for example, is seen as keeping people safe from natural threats and from each other, then the theories ought to be chosen for their prudential virtues. These would be features of a theory that indicate that believing it won't lead to people getting hurt. Given this kind of goal, the re-sponsibility of the sciences is a prudential responsibility, adopting theories that will not endanger people's safety. A prudential vir-tue might be that the theory is free of any descriptions of the re-lease of large amounts of energy. For reasons of safety, that is, one might want to avoid a theory like $E = mc^2$. Conversely, it might be advisable to believe in a theory like the greenhouse effect and global warming, just to be on the safe side. If a theory predicts a preventable disaster, it makes good prudential sense to adopt the theory and implement the preventive measures, whether the the-ory is true or not. Better safe than sorry. Pascal was thinking along these lines when he decided to believe in the theory that

God exists. Think of it this way, he argued: Believing in God is safer than not, for if you believe in God and he exists, you have invested in a pleasant afterlife. If you believe in God and he doesn't exist, what's the loss? On the other hand, if you *don't* believe in God and he exists, you're in for trouble, and if he doesn't exist, you gain little in terms of making your life safe and pleasant. Clearly, predicting catastrophe for nonbelievers is a prudential virtue.

Of course, there are other standards by which to evaluate theories. Some of the features of theories will stand out as pragmatic virtues, properties that make the theory easier to work with or more useful. All other things being equal, for example, it is wiser to go with a simpler theory than with a more complicated one. If nothing else it will be easier to work with and less likely to incur mistakes in calculation or in reasoning. There are good practical reasons, in other words, to choose a simple, streamlined theory.

Theories may also have aesthetic virtues. Because of its symmetry or the elegance of its models, a theory could be appealing to the eye. These features just make it look good. The feature of being simple could be doing double duty here. It could be both a pragmatic and an aesthetic virtue.

There are also psychological virtues. If a theory has the property of giving us a feeling of understanding, of explaining something so that it now makes sense to us, it has psychological virtue. Astrological theories are like this in that, once you believe them, a lot of things are explained. My theory about the missing Sunday paper has similar virtue. I feel better now that I have an explanation for the missing newspapers. Before, it was so mysterious as to make me a bit uncomfortable.

All of these kinds of features of theories are nice, but they are not what we are interested in. Insofar as the aim of science is to deliver a true (though not necessarily safe, practical, beautiful, or satisfying) account of the world, it must focus on truth-conducive features. A theory with these features is more likely to be true

35

than a theory without them. These are the features we are interested in to answer the question, Why believe what science tells us about the world? It's not that the other kinds of virtues are simply unimportant. They are unimportant only for the particular goal of assessing the trustworthiness of science. Whether it's practical or not, safe or not, or beautiful or not, is it true? Any feature that supports an affirmative answer to this question is a truth-conducive virtue. From here on, truth-conducive virtues will be referred to as simply virtues.

Our project is to articulate the virtues. Within science, justification of a theory is the presentation of virtues. This is the process that wears away the label "hypothesis."

Some of the virtues mentioned earlier under other aims may serve as well to indicate truth. Simplicity, for example is a pragmatic virtue; it may also be a truth-conducive virtue. Being able to explain observed phenomena, for another example, is a psychological virtue; it too may be a truth-conducive virtue. We'll have to see. In other words, it remains to be demonstrated that these features of a theory indicate that the theory is likely to be true.

INTERNAL VIRTUES

Here is a description of important internal features of theories that are truth-conducive, that is, of internal virtues. This list is not above revision. There may be good reason in future to add to it or delete from it. It is, however, a good beginning for justifying theories.

1. Entrenchment

"Entrenchment" is not a very flattering name for a standard we regularly apply to assess the warrant to believe that a theory is true. A theory is more believable if it is compatible with other, well-established beliefs about the world. In other words, a theory should be plausible, given what we already know. This requirement of initial plausibility is an impediment to wild and outlandish hypotheses that challenge the bulk of what is believed to be true of the world.

This property of entrenchment comes in degrees. A theory could fit in with other, background theories in the rather minimalist sense of simply being consistent with them. This could be pretty easy since a theory that talks about something entirely new will be consistent with all the old, entrenched theories no matter what it says. Where there is no contact there can be no contradiction. The theory of my neighbor the thief, for example, is perfectly consistent with the special theory of relativity, the theory of evolution, and all those other well-established beliefs about the world, and in that sense it fits in. It is a lonely fit, though, and a stricter level of entrenchment would have the theory be derivable from some other theories. The background knowledge should be not only compatible but also cooperative. In this sense, the fit, and the entrenchment, can be better or worse.

Entrenchment is a principle of conservatism, and it is clearly not a rule that is or should be followed with strict rigidity. Sometimes well-entrenched theories are refuted and wild and crazy theories turn out to be true. And sometimes calm, responsive, sincere witnesses are lying through their teeth. Einstein's special theory of relativity, claiming that measurements of mass and the passage of time depend on the motion of the observer, certainly lacks the feature of entrenchment, yet there is good warrant to believe the theory is true. Clearly, there are circumstances in which it is appropriate to relax or even ignore the standard of entrenchment. Clearly, or we would still think that everything is made of the elements earth, air, fire, and water.

Other things being equal, though, it is best to choose the theory that is compatible with the well-established background theories of the world. This way, the justification enjoyed by the entrenched beliefs is shared with the newcomer.

Entrenchment is the first of the virtues to be looked at closely, and it is the first indication that it is not going to be easy or straightforward to add up the score on a theory's virtues. The standard of entrenchment, like the other standards we will discuss, must be imposed with discretion, sometimes taken seriously but other times disregarded. The hard part will be in knowing, in the case of entrenchment, when conservatism is called for and when it is not. The answer will depend on the status of the other virtues. It must be all things considered.

2. Explanatory cooperation

Explanatory cooperation should really be an aspect of entrenchment, but the activity of explanation has a very high profile in science and is thereby deserving of its own status as a virtue. Explanatory cooperation means cooperation with other theories. It is the property of explaining, in part, why things are the way other theories say they are. Recall the example from the courtroom. Testimony that it was dark at the scene of the crime explains other testimony that no one was seen in the back seat of the car. The description of the crime starts to make sense now. The picture fits together and is therefore more likely to be true than if it were disjoint. In a scientific example, the theory that the boats used by common Greeks in the Bronze Age were small and unable to hold much cargo or travel far from the shore would explain, in part, why, as theory has it, fish was but a minimal part of diet in Bronze Age Greece. This explanatory cooperation makes both theories look good. It's not that justification flows from one to the other but that the explanatory link gives them mutual support.

This is different from explaining observed phenomena. It is different, for example, from the theory that fish was only a small part of the ancient diet explaining why only very few fish bones are found in excavations of Bronze Age sites. Being able to explain the observed phenomena is a very important feature for a theory to have, but it is an external feature. Its time will come.

3. Testability

Testability is a crucial aspect of science. A theory must make some predictions of observations that can be checked to see that the theory makes contact with the world. Sometimes this requirement of testability is identified as the essential ingredient that makes science scientific.

Testability is not really a truth-conducive virtue since being amenable to a test is different from passing the test. It is, though, a requirement for responsible justification of theories. A theory that lacks the feature of testability is precluded from evaluation for external virtues. That is, it can't have any external virtues. Because of its important role as a necessary prerequisite, the language can be stretched just a little to include testability as a virtue.

Being testable does not mean the same thing as being tested or as having passed tests. These features are clearly important, but they are external virtues and their time too will come. What testability does mean is that the theory is not insular and self-protective. It puts its credibility on the line by making predictions that are not guaranteed to be true, by being logical tautologies, say. This is a mark of responsibility. If the theory is false it will be exposed as false. Without verifiable predictions, the theory is left to its own proclamations of truth just as a witness on the stand who says nothing that can be independently checked must be taken at her word.

39

Sometimes the feature of testability is described using the concept of falsifiability or refutability. This is, of course, distinct from being falsified or refuted. It means simply making predictions that are not predetermined to be right, predictions whose outcome depends on how the world is. In this way, the testing of the predictions will be meaningful and is more likely to expose weaknesses of the theory. Again, if the theory is false, this is how we will find out.

The more precise a prediction, the more informative will be the testing, and so precision is an aspect of testability. A vague prediction could too easily be counted as true even if the theory that sponsors it is false. The smaller the target, that is, the more one's skill is tested. Theories of astrology tend to fail on this evaluation. Based on the arrangement of stars and planets at someone's birth, the theory predicts such things as a tendency toward generosity or a tendency toward seriousness. How can you test things so nebulous as tendencies and so imprecise as seriousness? Hitting such large and unclear targets is no real test of the accuracy of the theory. Testability, in other words, requires not only observable predictions but precise predictions as well. If I propose a theory, for one more brief example, that predicts that some sort of earthquake will occur sometime in the next hundred years, the prediction is observable and will quite likely be true. It won't however, be much of a test of the theory. If this is the extent of its predicting, the theory is not falsifiable enough to be taken seriously.

In sum, then, testability is an important prerequisite for a good scientific theory. It is a feature of the structure and content of the theory which indicates that it is amenable to testing against the observations. It is an internal virtue which is necessary for there to be any external virtues. The theory must first make predictions and explanations before these can be compared with observation. The predictions must be neither predetermined, as are logical and mathematical claims, nor indeterminate, as are very vague claims. How the theory fares in its contact with observations, that is, whether the precise predictions are true or false, is then an external feature.

4. Generality

General claims about the world are more interesting than specific claims, but that doesn't mean they are more likely to be true. There is, in fact, a kind of trade-off in that the more sweeping a statement, the more risk it takes. On the other hand, if there is some basic unity to the world, some small number of fundamental parts and operating principles, then only very general theories can capture these basic truths. With the assumption of unity, the feature of generality is a truth-conducive virtue. Without the assumption, the values of generality are pragmatic, aesthetic, and psychological in that generality is an aid in organizing our understanding.

Besides being a clue that a theory is getting at the deeper truths about the world, generality is a valuable feature because it enhances a theory's testability. If the theory generalizes over time such that its description is applicable at all (or many) times, then the theory is amenable to repeatable tests. Relevant experiments can be performed more than once with the expectation of the same results, thus making the tests more reliable and more accountable. This is a virtue. If a theory generalizes over many kinds of things, then it is amenable to a variety of tests. The greater diversity in the testing sample, the more at risk is the theory and hence the more significant is the testing. Generality, in the sense of describing a large variety of things, enhances falsifiability. If, for example, a theory describes some aspect of the characteristics of all electrons and not just the ones in this room or just the ones in hydrogen atoms, then the theory

41

has the obligation to predict correctly the electron-related behavior in all conditions, not just the limited circumstances in this room or in hydrogen. The theory is in this sense more open to testing.

5. *Simplicity*

Simplicity is a popular choice for an internal virtue of a theory. A good theory should be simple, or, at least, the simpler the theory the better. This approach to theory assessment is sometimes referred to as using Ockham's razor, after the medieval philosopher whose methodological advice was to shave theory clean of all unnecessary complication. Make it as simple and uncluttered as possible. Not only is the streamlined theory more likely to be true; it is also much easier to work with and more rewardingly elegant to present.

There are two points to clear up before simplicity can be used as a truth-conducive virtue. They are exactly the two questions that must be asked of any feature that is cited as indicating that a theory is true: How can the feature be evaluated? Does it really indicate that the theory is likely to be true?

How can we tell that a theory is simple, or how can we compare two theories to determine which is simpler? Sometimes it's easy. If the theory is mathematical, it can be rather straightforward to assess the degree of complication of the equations, whether they are linear or quadratic functions or perhaps exponentials and hyperbolic cosines. Also relevant to the measure of simplicity of a mathematical statement is the number of variables and constants. The more of these participants needed in the formula, the more features needed in the description of the world, and hence the world seems more complicated.

Even without the formulas and equations, it is easy to see how mathematical simplicity can be determined and used as a factor in theory choice. Consider the example of a scientific experiment

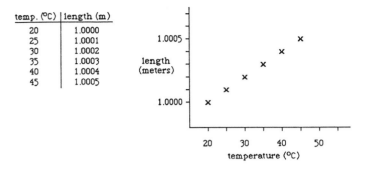

temp. (°C)	length (m)
20	1.0000
25	1.0001
30	1.0002
35	1.0003
40	1.0004
45	1.0005

Figure 2.1. The data of measuring the length of a brass bar at various temperatures, presented on both a table (*left*) and a graph (*right*).

to understand the relationship between the temperature of a brass bar and its length. At room temperature (20° C) the bar is one meter long (1.000 meters, to be precise). The bar is heated and its length is measured at five-degree temperature intervals, that is, at temperature 25° C, again at 30° C, and so on. The results of the length-versus-temperature measurement are tabulated and plotted in Figure 2.1. At this point the observed phenomenon has been carefully described. Theorizing begins with the projection of the general relation between length and temperature, that is, the description of what the length would be at any temperature, whether it's one of the ones measured or above these or in-between. A theory would specify how to draw a line on the graph that passes through the measured points and extends the information to intermediate and higher-temperature points. Given the finite number of measurements, the finite number of data points on the graph, that this and every other experiment are limited to, there will be lots of lines that pass through the data points. There will be lots of theories that are compatible with the observations. These theories have equal scores with respect to external virtues, and this is exactly why internal virtues are so important. If an appeal to the observations is indecisive between

43

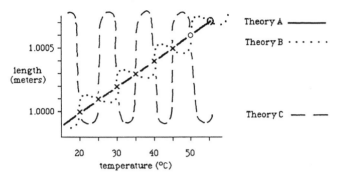

Figure 2.2. Three possible theories, *A*, *B*, and *C*, which agree on all the data, both past (*x*) and future (*o*), but disagree on the properties of the brass bar.

theories, the choice between the theories must be made not simply on the basis that they account for the phenomena but on *how* they account for the phenomena.

With the example of the heated bar, a few of the many possible theories, all of which are compatible with the data, are sketched in Figure 2.2. The points on the graph don't decide between theories *A*, *B*, and *C*. Not even tests of predictions of what the length will be when the temperature is 50° C will be decisive. But an appeal to simplicity rules in favor of theory *A*, the straight line. As Ockham might say, why think that the unmeasured lengths oscillate wildly back and forth when the length we chose to measure increases so clearly in a straight line? In this case it is easy to see which is the simplest theory, and it is easy to see that its simplicity is counted as a virtue.

Usually, though, evaluating simplicity is not this easy. Interesting theories, those that are challenging and important to justify, are never just lines through data points on a graph. What about the theory of evolution, for example? Is it simple? More to the point of decision making, is the theory of evolution simpler than the theory of special creation? It's not just that this question is

44

difficult to answer. It's difficult to know how to begin the evaluation to come up with an answer. And this is typical in assessing the simplicity of theories. It's unclear, to cite another example, that Einstein's special theory of relativity, with its description of time and mass as being dependent on the observer's motion, is simpler than Newton's mechanical account of the world.

Furthermore, the measure of simplicity of a theory will be influenced by the language and concepts used by the evaluator to describe the world. Different languages and theories will describe the same situation with different numbers of things. For example, there is a large variety of elemental building blocks in the world. There are hydrogen, helium, carbon, xenon, and so on. In the context of the atomic theory, though, these are all the same kind of thing. They are all atoms. Employment of the atomic language, then, shrinks the number of kinds of entities to one, and if simplicity is to be judged on the basis of number of entities, it will be a feature of the choice of language rather than the content of a theory. The measure of simplicity, in other words, is not an independent, objective standard but is dependent on the linguistic and theoretical context. This is true even in the mathematical case. The appearance of the graphed data and theory, as well as the degree of complication of the descriptive formula, depends on the coordinate system used, that is, on the choice of mathematical language. The straight line in the heated-bar example is the simple result of our choice of Cartesian coordinates. Plot the same data in polar coordinates and the simplicity disappears.

Simplicity, in other words, is not an unambiguous or easy thing to evaluate. It is sensitive to the choice of language of description instead of being stably anchored to the content of the theory. This not only makes it hard to assess but makes its link to truth questionable.

What does simplicity have to do with truth? Worries aside about the difficulties of evaluation, why should simplicity in a theory be taken as a sign that the theory is likely to be true? This is not to suggest that simplicity is not a truth-conducive virtue, but that claim does bear the burden of proof. Simplicity is clearly a pragmatic virtue, and for that reason it is a good thing to strive for. But we have yet to see the connection between being simple and being true.

This quick survey of some of the important and popular internal virtues has shown that none of them can be evaluated in a straightforward way. The answer we get as to how virtuous is a theory depends on our other theoretical preconceptions. This is clearly true of entrenchment and explanatory cooperation, and it is true as well of simplicity. Different people, with different beliefs about the world, will see different theories as having different internal virtues. This clearly limits the value of these virtues as objective criteria for justifying and choosing theories. It indicates, as well, an aspect of the self-regulation of science mentioned in the Introduction. If these standards of justification depend on the background language and theories of science, then the application of the standards is a task for the experts who speak the language and understand the theories. It is not just a scientist–nonscientist separation that is important here. The users and believers of theory *A* may honestly see their claims as simpler while the users and believers of theory *B* judge their own claims as simpler. If the judgment on virtues is theory-dependent, it seems the virtues cannot always be decisive between theories.

As further complication in the use of internal virtues to justify theories, note that there will always be conflicts among the virtues of a particular theory, and trade-offs must be made. There will be conflicts among internal virtues and conflicts with external virtues. Simplicity, for example, conflicts with generality. The speculative new theory of superstrings, which unifies the quantum descriptions of matter and the relativistic description of gravity, thus tipping the scale of generality, does so at the expense of describing the events of the world in twenty-six dimensions instead of the usual four (three spatial dimensions and time). The theory does well on generality but poorly on simplicity, and, indeed, no theory will win across the board. Nor is there any indication of which virtues are more important. There is no hierarchy to rank the importance of virtues so that if one infringes on another it is clear which has the higher authority. In this sense, the list of internal virtues is like the Bill of Rights. Neither specifies how to resolve conflicts among various rights or values, and so both must be administered with discretion.

This discretion involves the recognition that even though these features of theories are virtues they must sometimes be compromised. It may be justified to believe a theory that lacks particular virtues, justified because the theory is better on other virtues, possibly including external virtues. Obviously, it is unwise to be a slave to entrenchment, as this would stifle any change in science. It is nice, and probably justified, to think that some progress has been made since the theory of four elements, earth, air, fire, and water. Nor is it advisable to depend single-mindedly on simplicity, lest science become simpleminded. There are trade-offs to be made, both for other internal virtues and for external virtues.

EXTERNAL VIRTUES

External virtues are very important. They must be a heavily weighted part of the total score in justifying a theory. These are the features, after all, that indicate that the theory is compatible with experimental results, that is, that the theory is accountable to observations. These virtues will be only briefly described now because their importance warrants separate chapters.

1. Explanation

A good theory should be able to explain some observed phenomena. For example, the coincidence between the ocean tides and

47

the position and phase of the moon has been apparent for a long time. Newton's theory of universal gravity explains this coincidence by pointing out that the situation is an instance of a general law. The objects are the kinds of things that gravitationally pull on each other, and so it is to be expected that the pull of the moon would cause the water in the ocean to move. The explanation shows how the particular phenomenon fits in the general pictures of things, and this fit makes the picture seem accurate.

Even better, of course, is for the theory to explain lots of phenomena. The more pieces of the picture that can be shown to fit, the bigger the sample of explained phenomena, the more likely it is that the theory is true. Better still, a good theory should explain a wide variety of stuff. Instead of building a large sample by doing the same test over and over, a theory's truth is put more on the line by a broad sampling of tests. The fact that Newton's theory can explain not only the tides but also the orbits of the planets and the motion of falling objects weighs heavily in its favor, more so than the fact that it explains the tide on Monday, Tuesday, Wednesday, and so on.

It is important to note that there are other theories that explain the tides. The general theory of relativity is one, and I'm sure we could find or invent plenty of others. Here is one in the invented column. Maybe there are beautiful but shy sirens who live deep in the sea, sirens whose great love of the moon causes them to rise from the water when the moon is overhead. They rise and part the seas, forcing the water to swell up on the shores, thus explaining the relation between the moon and the tides. It's their shyness that has kept the sirens from being seen.

This is a stupid theory, and it does not warrant anyone's belief. But its failure cannot be blamed on a simple lack of external virtues. It does explain the tides, and with sufficient fixing up it could accommodate lots of other phenomena. It could be made compatible with the data. It must be excluded, then, on the basis of internal virtues. For example, it is not a very testable theory in that it can blame any failure to see the sirens on their convenient shyness.

The moral of the story is this: As important as it is for a theory to explain the phenomena, being an explanation does not decide unequivocally in favor of a theory. It has to be a good explanation, even the best explanation, and this measure of goodness is an internal feature.

2. Testing and confirmation

Testing is not strictly distinct from explanation, since it too is done by drawing observational consequences from the theory and seeing that they fit the facts. The difference is that testing is usually accomplished by looking at new evidence as predicted by the theory, whereas explanation is usually of phenomena already observed. The emphasis in testing is not so much on *why* something happens as simply on the fact *that* it happens according to prediction. Newton's theory of gravity, for example, not only explains the observed coincidence of the tides and the position of the moon but can predict occurrences of eclipses. These precise predictions are exactly what counts toward falsifiability. If an eclipse occurs as predicted, that is a virtue of the theory and the event is counted toward confirmation or corroboration of the theory. If the eclipse does not occur, that's a flaw of the theory and the event is counted toward falsification or refutation.

This scheme accurately captures the basics, but only the basics, of the testing of theories in science today. The current theory of the sun, for example, describes a furnace of nuclear reactions in the solar interior, producing, among other things, a lot of neutrinos. The theory predicts, then, that some of the neutrinos ought to be detectable on the earth. A productive step toward answering the question "Is this theory true?" is to look for the neutrinos, that is, to see if the prediction is true. This is an experiment in progress, and we will hear more about it later, not only in this book but in the science news.

The external virtues warrant and will get a much more careful analysis than they got in this chapter. They deserve this care because they are so important. Accountability to observation is much of what intuition says is scientific about science. They deserve a deeper look also because they are significantly more complicated than has been indicated so far.

The closer look at external virtues of theories will eventually lead to the demonstration that the dichotomy between internal and external features is untenable. What look like external features, it will turn out, really aren't. Insofar as features are relevant to justifying theories, they are internal. What will happen, then, is that the lists of internal and external virtues will be merged and expanded. In all discussions of virtues, it is important to be keen to the following questions: How can the feature be evaluated? Does it indicate that the theory is likely to be true? How is it to be used with discretion, and how are we to make trade-offs if this virtue conflicts with others?

50

3

EXPLANATION

P ROVIDING explanations of the phenomena we experience is
a major accomplishment of science. A theory should be good
for making sense of the world, for explaining why things happen
as they do, if it is good for anything at all. Scientific curiosity begins with why questions. Why does this happen? Why are things
this way? Scientific explanations are answers to why questions.
Explanations are an accomplishment, and they are worth working for because they enhance our understanding of the world.

Explanations, in other words, are desirable in themselves as
valuable ends. You want your theories to explain things because
that's what you want, just as you might buy shiny apples because
shininess is what you want. Explanation is a desirable end, but it
is also a means to something else. It is useful as an indicator of
truth, as shininess might be an indicator of freshness and good
taste. If an apple is shiny it is likely to be fresh, so even if shininess for shininess' sake didn't matter, shiny apples would be desirable for taste's sake. Similarly, it is a good thing for a theory to
be explanatory, not just for explanation's sake but for truth's sake
as well. A theory that explains phenomena is more likely to be
true than one that doesn't. That, at least, is the claim if explanation is an external, truth-conducive virtue. Clearly, though, it
cannot be quite so simple, since the caloric theory explains why
things expand when they are heated, the theory of oceanic sirens
explains the tides, and the theory about scraping the skin of old,
dried-out rubber off by rubbing a balloon on your hair explains
why the balloon then sticks to the wall. All these theories explain
things, and yet they all are false. Nonetheless, there does seem to

be something to the link between explanation and truth, and the way to find out just what it is is to analyze carefully just what it is to be a scientific explanation.

The emphasis here is on explaining observable phenomena, not on explaining other theories. This latter accomplishment is surely an interesting and valuable role that theories have, but it is an internal affair and one we have already discussed. Now we turn to external features. The theoretical claim that protons have a positive electric charge, as an example of the distinction between internal and external explanation, explains what holds electrons in their atomic orbits. It cooperates with other theories to explain the structure of atoms, objects that are never manifested in perception. No one would have asked "Why don't the electrons fly out of the atom?" had there been no atomic theory. But the charge on the proton also explains why some of the vapor trails in a cloud chamber curve to the right when the chamber is situated between the poles of a magnet but none of the tracks curve at all when the magnet is far removed. This is an event in the world of manifest phenomena. It is something observed, and explaining why it happens is part of what is meant by making contact with the world and being compatible with the data. It is an external feature if anything is. The question is, Is this external feature of explaining observable phenomena a truth-conducive virtue? We answer that by dissecting the process of explanation, first as it is portrayed in the standard model, and then with some amendments.

THE STANDARD MODEL

Any claim to be describing the form of scientific explanation is best understood not as presenting the single form into which all good explanations actually fit but as outlining a paradigm, an exemplar to which most good explanations are similar. It is a core concept with variations and extensions as particular circumstances warrant. It is the central idea from which actual explanations are more or less divergent and about which actual cases are centrally located.

The standard model of this central idea is called by philosophers of science the **covering-law model** of explanation. Its initial formulation is the work of Carl Hempel (listed in the Suggested Reading). It focuses on the form rather than on the contents of an explanation and thus follows the motivation of its originators to be generally applicable to all sciences. The pattern of a good explanation will be of how it is put rather than what it says, thus avoiding specifics of content which would draw in reference to one science or another. This goal of a general, formal characterization is a noble intention, but we will have to see how far it can go.

A phenomenon is explained by demonstration that it is an instance of a law of nature. This is the basis of the covering-law model; an event is explained when we articulate the law that covers it. This happens in everyday explanation as well as in science, and an everyday example is the best kind to start with. Suppose someone with the good fortune to be unfamiliar with automobiles and traffic asks why the cars are stopping before crossing the intersection. The why question solicits an explanation, and the explanation goes something like this. There is a stop sign on the corner, and everyone must stop at a stop sign. It's the law. The similarity between this example and laws of nature and scientific explanation is only of the form of presentation, citing a general law to cover a specific instance, and not at all of content. Traffic laws have an author and are imposed by us; natural laws have neither of these features.

There are two components of this and other good explanations. It specifies the relevant conditions of the particular case, that is, the properties of the event that are important to its happening as it does. In this case it is the presence of a stop sign on the corner. Furthermore, the explanation specifies the relevant law that relates the conditions of the case to the phenomenon in question. It's the law that cars must stop where there is a stop sign. This law, in other words, covers this case. So, once the law is understood and the conditions are recognized, the event is to be expected. Had the law and conditions been known, the event could have been predicted. This kind of understanding is what is expected of a good explanation, and the major achievement is the finding of the relevant law.

This model seems to fit nicely onto a lot of important examples of scientific explanations. Anyone who has spun anything like a ball on a string or a little brother by his ankles knows that revolving objects will fly off, away from their orbital center, unless there is some active force holding them in. If the string breaks, the ball flies off. If you let go of your brother's ankles, he's in the bushes. Well, the moon is revolving around the earth at a pretty good clip. Why doesn't the moon fly off? There's no string, it's not set in a rigid, crystalline sphere, and nobody has it by the ankles. This deserves an explanation. And the explanation is simple: gravity. The moon doesn't fly off because of gravity. This compact explanation can be magnified to reveal its vital parts. The appeal to gravity is an appeal to a law of nature, namely, the law of universal gravitation, to the effect that any two massive objects exert a mutual force of attraction. The relevant conditions in the earth–moon case, therefore, are simply that both are massive bodies. Given the law and the conditions, the event of an attractive force on the moon is to be expected; the particular phenomenon is covered by the law.

This is not to say that all good scientific explanations are in this form of explicit appeal to law and conditions. Often the law or the conditions are implicit, and the claim of the covering-law model is that good scientific explanations can be put into this form or something very much like it.

Once explained, a phenomenon is to be expected. This is the contribution to understanding. This expectation does not mean simply that the law is subtly evocative or suggestive of the phenomenon or that it puts you in the right mood to expect the phenomenon. There is a stricter connection between the law and the event in question, one that forces the expectation and removes it from subjective control. The statement of the law and the relevant conditions must entail the statement of the event. With stop signs, for example, the two claims, "everyone must stop at a stop sign" and "we are at a stop sign," logically entail the description of the event, "we must stop." The law is not merely a suggestion of the event; it is a logical requirement.

With this strong connection between law and event, the covering-law model is also referred to as the **deductive-nomological model** of explanation, or simply the D-N model. The word "nomological" means to pertain to laws, and **deductive** is there to describe the strict logical connection between the premises of the explanatory argument, that is, the statement of the law and relevant conditions, and the conclusion, the description of the phenomenon to be explained. It is a necessary connection in the sense that, if the law is true and if the conditions hold, the event must occur. This is the warrant for expectation of the event. This is how the covering law covers.

Following this model, an explanation is presented as an argument with premises and conclusion. Often some of the premises are not mentioned, but a full presentation of the explanation should be able to expose these and reveal a complete deductive argument. An episode of explanation typically begins with a why question. Why does event *E* happen? Or, why does it happen in the particular way that it does? We've already asked why the moon stays in its orbit around the earth. Such questions initiate the scientific process of inquiry. Why is the sky blue? Why are there so few fish bones found at the sites of Greek Bronze Age settlements? Answers to these questions, that is, explanations, are presented as an argument: It's because,

if law L (or laws L_1, L_2, . . .) is true
and conditions C_1, C_2, C_3, . . . hold,
then event E must be true.

To give an explanation of event E is to specify the laws L and the conditions C. Laws are recognizable as being general theories, whereas conditions are specific, observable features of the situation.

The moon example fits this form of presentation as an argument. Event E in this case is that there is an attractive force on the moon. It could even be a quantified statement regarding how much force (F) is needed to hold the moon in orbit. Only one law L is required, and it too can be given a quantified form since any two massive objects (with masses m_1 and m_2) separated by a distance (r) will exert a mutual force of attraction of magnitude

$$F = G \frac{m_1 \, m_2}{r^2}.$$

G is the gravitational constant to reconcile our units of mass and distance with our units of force. The particular conditions of this case are: C_1, the earth is massive (specifically, its mass is m_1); C_2, the moon is massive (m_2); and C_3, the distance between the earth and the moon is finite (r). Together, L, C_1, C_2, and C_3 deductively imply E. Set up as an explicit argument,

if L
and C_1, C_2, C_3,
then E.

By covering-law standards, this is the paradigm form of explanation, and recognition of a good explanation is based on its closeness to this form rather than on any specifics of content.

A richer sampling of examples of explanations and apparent explanations will raise indications that this formal analysis is insufficient. (Remember that the project here is to characterize explanation properly so that we can clearly tell when a theory is explanatory and so we can see what it is about being explanatory that indicates that the theory is likely to be true.) There will be examples of arguments that follow the D-N form but are not what we regard as good explanations. They do not, that is, answer the initiating why question, nor do they further the understanding of the phenomenon.

Consider again the law of universal gravitation, but start with a different question, asked in a different context. The setting is a geology rather than an astronomy class, and the question is, Why does the earth have a mass of 6×10^{24} kilograms? Is the earth hollow, or lead-filled, or what? If all that it takes to present a good explanation is an argument of law-plus-conditions that entail the event E (the earth has mass $m_1 = 6 \times 10^{24}$ kg), then the law of gravity will do. The earth has the mass that it does because of gravity. Expanded into proper D-N form, the alleged explanation goes like this:

L: Any two massive objects attract with force $F = G \dfrac{m_1 m_2}{r^2}$.

C_1: The moon has a mass (m_2).

C_2: The earth–moon separation is r.

C_3: There is an attractive force between earth and moon of magnitude F.

E: Therefore, the earth must have mass $m_1 = \dfrac{F r^2}{G m^2}$.

The form is valid. If the law is true and the conditions hold, the event must occur. This makes it seem as if an appeal to gravity and the circumstances of the moon explain the compositional fact about the mass of the earth, but that's silly. It's not a satisfying explanation at all. The lunar factors are irrelevant to the compositional characteristics of the earth, and the law of gravitation fails to address the cause of the phenomenon. It only mentions its effects.

The problem is that the formal analysis is symmetric but explanations are not. The mass of the earth causes its gravitational force and hence the attraction of the moon, so an appeal to masses works to explain a gravitational effect. Gravity, on the other hand, does not cause mass, and so the law of gravity does not explain the existence or magnitude of masses. But the logical form of D-N explanation is blind to this asymmetry. It licenses as good explanations such backward episodes as, Why is there a stop sign on this corner? Because we stopped here. It fits the pattern if there is a law to the effect that the only place one must stop is at a stop sign, and the condition that we must stop. These entail the conclusion that there is a stop sign present. The sign is there because we stop. No, that's backward. We stop because the sign is there, and a proper explanation of why the sign is there must cite the city's efforts at traffic safety.

These examples indicate that, even if the deductive-nomological form is a necessary feature of a good explanation, it cannot be sufficient. There must be additional criteria, most likely dealing with factors of causal relevance. These factors must be based on an understanding of the content of the theory cited in the explanation and of other theories that indicate what is relevant about the conditions. A theory informs us, for example, that mass causes gravity but not vice versa.

Another kind of counterexample to the D-N model indicates the same thing. Back in the astronomy class, we can ask the same question about the moon's attraction to the earth. The phenomenon to be explained is E, the moon's attraction to the earth, and the D-N form requires only that this be shown to be an instance of a regularity. E must be deduced from a general law and the relevant conditions, like this: C, the moon has craters. (This is true.) L, anything with craters is attracted to the earth. (This is true too.) L and C deductively entail E, so this explanation, that the moon doesn't fly off because of craters, has the proper D-N form. It is not, however, a good explanation. It's not even close. That's because craters, as far as we know, are irrelevant to forces of attraction. The generalization about craters is not even a law since it doesn't pick out a natural kind of thing and associate it with a particular effect. "Cratered thing," in other words, does not pick out a natural kind, at least not with respect to forces.

Thus the covering-law model of explanation includes a problem of distinguishing laws from accidental generalizations. This cannot be done simply on the basis of the form of the statement, that is, on its generality, since "all cratered things attract" is as general as "all massive things gravitate." The distinction must attend to the content of the claim to determine that the property that identifies the kind (for example, being massive) is relevant to the effect. Determination of what counts as a law, in other words, will depend on our theoretical background to guide the decision as to what is relevant and what is not. The theories we hold, then, influence what counts as an explanation. An evaluation of this external virtue, that is, will be influenced by an internal reference to other theories.

This need for determining relevance factors is shared by all explanations, even the stop-sign case. If, as in the beginning, the event in question is someone stopping at the intersection, the explanation, as allowed by the D-N form, could cite the following generalization: Everyone must stop when there is a sign that has an additional, rectangular sign below it that says "4-way." This is a true generalization (assuming that only stop signs, though not necessarily all stop signs, bear this addition), but it is irrelevant. It misses the real law by missing the relevant factor for why people stop. They do not stop in virtue of its being a four-way sign. They stop in virtue of its being a stop sign.

This account of the covering-law model of explanation and its problems has ignored several important issues regarding its adequacy. For example, what about cases of statistical laws in which the relevant conditions don't always bring about the phenomenon? The event of Sarah's death is explained by her having leukemia. But not every case of leukemia is fatal, so why did it happen *this* time? And there are other variations from the standard model that haven't been and won't be developed here, variations that deserve the attention of a thorough analysis of explanation. This brief account, though, is sufficient to raise the main issue of explanation as a truth-conducive virtue of theories. If being a good explanation of phenomena is to function as a truth-conducive virtue, it must be a feature we can evaluate. What *is* a good explanation? Analysis so far indicates that this feature cannot be evaluated without reference to some theories. It must be clear that the law and the conditions are describing *relevant* factors and not just circumstantial properties like having craters or a four-way sign. Any appeal to a determination of relevance will involve reference to other theories, theories that describe the fundamental workings and causal connections that make things happen as they do. Judgments of relevance, in other words, are internal judgments, getting their authority not from experience directly but from theory.

The moral of the story is that an adequate account of scientific explanation requires more than the formal guidelines provided by the covering-law model. If the covering-law model is to characterize the basic idea of explanation, it will need some amendments.

AMENDMENTS TO THE COVERING-LAW MODEL

One suggestion is to add to the formal covering-law model a measure of unification. An explanation must reveal a unity in nature, and it must be fruitful in the sense of covering a maximum number of phenomena with a minimum number of laws. This, after all, is the accomplishment of understanding, finding the most basic pattern in the events and structure of the world. The more fruitful the explanation, the better the explanation. When there are competing explanations, as between gravity and craters as the explanation of the moon's attraction to the earth, or between stop sign and four-way sign as explanation of the traffic stopping at the intersection, one should always opt for the more fruitful. In this way, covering-law-plus-fruitfulness will weed out the silly and irrelevant attempts at explanation.

This new scheme avoids the previous problem of irrelevant generalizations masquerading as explanations, that is, of true generalizations which, with true conditions, entail the event in question but which do so on the basis of irrelevant features. The key is that accidental associations, as between craters and an attractive force, won't be extendable to other, similar situations. The accidental generalizations won't be projectable. They won't be fruitful and unifying as is required of genuinely explanatory laws. The alleged law relating cratered objects to an attractive force could be used to explain only a limited number of cases of orbital forces, namely, those involving cratered objects like the moon, but the law relating massive objects to the force will serve all cases of moons, planets, satellites, and comets. Even without a precise indication of how to measure fruitfulness, this case is clear. The law of gravity wins, and it is to that law that one should appeal to explain the moon's behavior. Being able to unify so many different phenomena under the same law must be indicative that the law is getting at the relevant factor, mass not craters, for the effect.

The same analysis works in agreement with intuitions about the four-way-sign case. The generalization about stopping at signs with a four-way tag will cover some cases of stopping at intersections, but the generalization about stop signs will cover all those and more. This feature of fruitfulness is taken to be a sign that the generalization is describing the operative factor in the law and hasn't simply happened on a circumstantially true association.

This appeal to unification and fruitfulness will help to avoid the silly cases of explanation that result from the overly generous symmetry of the D-N model. Unification adds the asymmetric factor into the account and blocks the sort of role switching between condition and event that allowed the moon's gravity to explain the earth's mass. The structure of the explanation will be more fruitful one way than another, thus identifying one case as the proper explanation to the exclusion of the other. All astronomical orbital systems, for example, can be explained by the law of gravity and the specific conditions of the masses of the participating objects. But not all masses can be explained by their participation in an orbital system, that is, by the law of gravity and the specific conditions of the force of attraction. This is simply because not all masses are in orbital systems. Some are isolated in deep space. Explaining the masses of these objects cannot be in terms of the forces on orbiting companions but must invoke some other generalization, thus compromising the unity in the picture of the world. The unification shown by the law of gravitation explaining all cases of orbital forces shows that the law picks out the relevant factor to bring about the phenomena and hence that it is the appropriate explanation of this kind of event. The failure of the same law to achieve the same unification in explaining the masses of objects shows that it misses the relevant point and is inappropriate as an explanation.

This example, though, also shows that the concepts of unification and fruitfulness are not as clear as they could be. In declaring the gravity-explains-orbital-attraction case to be more fruitful and of greater unification than the gravity-explains-mass case, we were operating largely on intuition and without a clearly articulated way to measure these features. Fruitfulness demands a trade-off between many phenomena explained for a few theories cited, but what is the optimal formula? If three theories buy you a hundred explanations, is that better than two theories that buy fifty? Is it simply brute numbers of phenomena explained, or is there extra credit for a large *variety* of phenomena explained? And how can you tell one variety from another? These are the kinds of questions that must be answered to clarify the measurement of unification and to make it useful as a criterion of explanation.

An alternative approach to amending the covering-law model of explanation is to require that an explanation include explicitly *causal* laws or causal theories of some kind. This may actually be getting at the same idea as unification in that a more unifying explanation may indicate that the law has focused on the essential causal ingredient. Though this link has been suggested in the account of unification, it has yet to be proven.

The specific focus on the causal factor in explanation represents a change of emphasis from the standard covering-law model. The goal of explanation is no longer to show that the phenomenon is to be expected. It is rather to show what brought the phenomenon about. Instead of explaining why an event happened by saying that these sorts of things always happen this way, that is, by showing it to be an instance of a regularity, the idea now is to explain the event by identifying its specific cause. Why did the dinosaurs all die? Because of a huge meteorite. The meteorite caused their deaths. Why are there tides in the ocean? Because of the moon. The moon's gravitational force causes the tides.

The requirement of an explicitly causal explanation can prevent the problems of irrelevant explanations and the silly symmetries. Since the craters on the moon don't cause the force of attraction to the earth, and no one claims that they do, the generalization about cratered objects and attraction to the earth is no explanation of the phenomenon. Similarly, the four-way sign is not what causes drivers to stop, and so even though they stop every time there is a four-way sign, this does not explain their actions. If there is no mention of the cause, then there is no relevance as an explanation. In a similar way, since the moon's attraction doesn't cause the earth to have the particular mass that it does, the moon's attraction cannot be used to explain the earth's mass, and this attempted symmetry in the argument is blocked. Stopping at an intersection, furthermore, does not cause the placing of a stop sign, so the analysis catches up with intuition and rules, as we knew all along, that the fact that someone stops doesn't explain the fact that there is a sign. Since the causal connection is asymmetric, causal explanations are, as they should be, asymmetric.

Reference to causes looks good until we realize how difficult it can be to distinguish a genuinely causal account from an account that is not causal. How, for example, could we recognize the causal from the noncausal theories in the previous examples? That is, how do we know that craters don't cause attractive forces but masses do? The decision is founded on our theoretical background, that is, on our theories about the mechanics of gravitation. Recognition of a causal explanation, in other words, can only be done in a theoretical setting. It is clear that this characterization of explanation is no longer dependent only on the form of presentation. Different theoretical preparation will lead to different rulings on what is or is not a causal factor and hence different rulings on what is or is not really an explanation. Again, different theories in place will yield different scores on this external virtue of explanation. It is not simply a case of comparing a theory with the facts. Rather, it is a case of, under the influence of several theories, comparing a theoretical claim with the observed phenomena. In this sense, theories never confront the observed world alone but only in groups.

The source of the difficulty in identifying genuinely causal theories is that correlations between events, no matter how regular and frequent, do not necessarily report a causal connection between the events. There is a difference between two events or properties occurring together with high regularity and one event or property causing the other. Whenever the sun is high in the sky the hour hand on my clock points up high as well. But neither event is the cause of the other, though they are observed to be closely correlated. The question is, How do we tell the differ-

ence between mere constant conjunction like this and actual
causal connection, when constant conjunction is all we can detect? It is exactly this difficulty that the tobacco industry uses as
a foil against arguments that smoking causes cancer. Even if there
is a strong statistical correlation between incidents of smoking
and incidents of lung cancer, this does not by itself support the
conclusion that the smoking is the cause of the cancer. The answer is ambiguous as to which event causes the other, if there is
any causal connection at all, nor does it rule out the possibility of
there being a common cause, perhaps stress and anxiety, which
brings about both events. This is important because if an unmeasured third condition causes both, then not smoking will not be
an effective way of preventing cancer, just as resetting the hour
hand of my clock will not be an effective way of moving the sun
in the sky.

The moral of this story is by now familiar. Determination of
causal theories, and determination of proper explanation, is neither purely formal nor purely from observational evidence. It is
guided by theory, influenced, that is, by internal concerns.

EXPLANATION AND TRUTH

There is no disputing the claim that explanation is a good thing
in itself. It adds to our understanding of the world, and in making
the phenomena comprehensible it satisfies our scientific curiosity
and helps us deal with the world we inhabit by knowing what to
expect from it. It is valuable in this sense whether it is merely describing regularities or specifying actual causal relations. But does
successful explanation also indicate that the explanatory theory
is true? Is explanation a truth-conducive virtue?

Successful explanation certainly doesn't guarantee that the explaining theory is true. The fact that a theory can explain phenomena does not deductively prove that the theory is true. What

then is deductive about the deductive-nomological form of explanation? It is the entailment from law (and conditions) to the event. *If* the explaining theory is true (and the conditions hold), then the event in question must happen. *Assuming* the theory is true, the phenomenon is to be expected. This is clearly not a deductive proof that the theory is true, and to argue as if it is is to commit a classic logical fallacy. An argument of the form "the law (and conditions) entails the event" and "the event is real" so "the law must be true" is a fallacious argument. The law being true is a sufficient explanation for the observed event, but it is not necessary. There may be other ways to explain it, and these have not been ruled out. The alleged deductive proof of the theory is just as flawed as arguing that whenever the day is the fourth of July it must be summertime, and it is summertime so it must be the fourth of July. Crazy. It could be August. The form of this argument is identical to the form of a deductive argument that a theory that explains an observed phenomenon is true. It is a form that is always fallacious.

Explanation is not a deductive proof that the explanatory theory is true, but it might contribute to an **inductive** proof. The difference between deductive and inductive proofs is exactly the difference between certainty and likelihood. The suggestion then is that successful explanation increases the likelihood that the explanatory theory is true. This would make explanation a truth-conducive virtue, if it can be demonstrated.

Does explanation in the mold of the standard covering-law model indicate that the explanatory theory is likely to be true? Recall that the accomplishment of this form of explanation is to show that the phenomenon was to be expected. A good explanation makes good sense out of what had been puzzling. It is, as C. S. Peirce put it, "feeling the key turn in the lock." All of these criteria of good explanation, this feeling, things making sense, expectation, are psychological factors that depend on a person's interests, antecedent beliefs, and psychological preparation. None of them really depends on the truth. That satisfying feeling of understanding can come from a false belief as easily as from a true belief.

A theory can be explanatory, in other words, without being true. For example, the theory that my neighbor is stealing my Sunday paper explains why the paper is not on the porch as it used to be. It all makes sense now, and it's to be expected that the newspaper won't be there. It all makes sense, but it's false. Nonetheless, if I continue to think *as if* the theory is true, I continue to make sense of the phenomena in my world. I know not to go down to look for the paper. I know I have to go buy one at the market. Now that I understand the phenomenon I can deal with it. The explanation can fully serve its purpose even though its pivotal theory is false. The same is true of more interesting examples, that explanations can succeed in making sense of phenomena, in giving us the feeling of understanding and expectation, whether they are true or false. Thinking *as if* the stuff around us is made of atoms helps us understand why the stuff behaves as it does. There's no doubt that the world is *as if* there are atoms and *as if* the guy next door pinches the paper, but it is a separate issue altogether as to whether the world really *is* this way. The psychological function of explanation, turning the key in the lock, can be accomplished without the theory being true. Truth and explanation in this sense seem to be independent issues.

Things are different if the causal factor is added to the model of explanation. A theory can explain an event, in the sense of showing that the event is to be expected, even if the theory is false, so one can accept the claim that the theory explains the event without committing oneself to believing that the theory is true. On the other hand, nothing can be the cause of an event unless that thing really exists, that is, unless the theory about it as a cause is true. Acceptance of a causal account as explanation of a phenomenon carries with it a commitment to the truth of the causal theory. Fictions can help us make sense of observable phenomena, but fictions cannot cause real phenomena. Effective causes must be real causes.

The big idea here is to figure out whether a theory is likely to be true and to do this based on the evidence of the theory being able to explain observable phenomena. Different criteria for explanation naturally facilitate this inference to different degrees. If you know that the theory explains the phenomenon in the sense of showing that it is to be *expected,* you don't know that the theory is true. But if you know that some particular thing *caused* the phenomenon, then you also know that this thing really exists.

That's great, but how do you *know* that this particular thing is the cause of the phenomenon? A causal explanation is informative of the existence of the cause only if it is the *correct* causal account. The question then is how to tell the correct causal account from all the incorrect ones. The trouble with causal explanation as a truth-conducive virtue is not that it doesn't indicate truth (because it does indicate truth) but that it is hard to evaluate. It is hard to tell when we've got a genuinely causal explanation.

Here is a dilemma. Explanation construed as an accomplishment of expectation and understanding, that is, of psychological features, is easy to evaluate but is not indicative of truth. On the other hand, explanation construed as identification of the cause is indicative of truth but is too difficult to evaluate.

We shall have to work on this, and the work will proceed by developing a way to evaluate the goodness of a causal explanation. A causal theory will be judged in the context of other theories. The credibility of describing something as the cause of a phenomenon will be a credibility according to other theories. Thus causality and success as an explanation are evaluated by comparison to theories, and explanation as an external virtue, as a form of comparison between theory and data, is seen to have an internal component.

4

CONFIRMATION

THEORIES are pretty easy to come by. Even with the demand for genuinely explanatory theories, there are lots of competing claims about the unseen objects and events that genuinely explain the phenomena we see. But at most one of those theories is true. The urgent responsibility of science, then, is to weed out the bad from the good, that is, to expose the failures in false theories and to champion only those theories that have endured a process of testing that rewards the true and scorns the false. This means that, as much as is possible, a theory must be tested against the observable data. For all a theory's theoretical elegance and pragmatic virtue, there is no reasonable claim to accurate representation of the world until it confronts the world. You've got to do the experiments.

This requirement of empirical testing is probably one's first and most intuitive answer to the question of what makes science believable, and it is a good answer. If the theories claim to be about the world, then they should be accountable to information from the world. Confirmation by testing against observations is the most straightforward external event in science, and achieving a positive test is the most obvious of external virtues. But it is important to be clear on what is meant by "testing" a theory and by "comparison with the observable evidence." Theories usually describe objects and events that are not amenable to observation. That's what makes them interesting, but it is also what blocks us from simply and directly comparing what the theories say to what is observed in the world. Take, for example, the definitive theoretical claim that oxygen atoms have eight protons. Checking this is not a simple matter of looking and counting; rather, it is a matter of checking on the consequences of the theoretical claim by determining what effect the eight protons must have on the observable world. No one sees the protons or the oxygen atoms, but it might be easy to see some of the effects they have. There is, then, this kind of indirectness in any case of using observational evidence to test a theory. It is a complicating factor worth keeping an eye on when deciding what it means to confirm a theory and whether there is any reason to think that certainty is a reasonable expectation of science. How close can we get to certainty, if not all the way?

To say that explanatory theories are pretty easy to come by is to say that the activity of discovery of theories is not the place to apply strict standards of scientific responsibility. There are lots of ways that people can and have come up with theories to explain phenomena, by observation, sometimes meticulous, sometimes haphazard, or by thinking, comparing and combining other beliefs to see what they suggest, or even by accident as in a dream or hallucination. However a theory is discovered or created, the place to apply rigorous standards is in the subsequent activity of justification. This is where science gets scientific, and it is this context of justification that is the topic here. You can dream up your theory in any way you like, but it must be accountable to observational justification if it is to be taken seriously.

This means not only that a theory must be testable, the internal feature discussed in Chapter 2, but that it must also be tested. Acquiring and evaluating the internal feature of testability is, as a rule, an inexpensive process, requiring only a clear understanding of the theory and the time and energy to do some calculations or whatever reasoning is appropriate for drawing out the consequences and for understanding the causal efficacies of the theory. The theory that describes how the sun works – for example, that the source of energy is an ongoing process of nuclear fusion at the center – is clearly testable. The nuclear fusion must produce huge numbers of neutrinos, a species of electrically neutral elementary particles, and some of these neutrinos ought to find their way to the earth where they can be detected. If nuclear fusion is the energy source, the sun must be showering us not only with heat and light but with neutrinos as well. This straightforward and inexpensive reasoning proves the theory to be testable since the neutrinos are detectable. Determining whether or not the neutrinos are in fact detected is another matter. This is the external affair of promoting a testable theory to the status of being a tested theory. Establishing this external virtue is the expensive part of science, for now we have to build the equipment and run the experiments. Consulting the world seems always to cost more than consulting our own thoughts.

This process of testing is known optimistically as confirmation, and in a sense there is no rigid dichotomy between the activities of confirmation and explanation. As will become clear in the following sections, they follow the same logical form, which compares the consequences of theories to observations of the world and thus connects theories with the world by observing the effects that the theorized entities have on objects we can see. If the neutrinos are detected we will say that the neutrinos contribute to the confirmation of the solar theory and that the solar theory explains the shower of neutrinos. It is typical, though, to apply the concept of explanation to cover past observations of events, things we knew happened but wondered why, and reserve the concept of testing to cover future observations, predictions whose outcome is yet unknown. Explanation and confirmation are also distinguished by the pragmatics of the situation, namely, what the scientists happen to be interested in. If it is the phenomenon itself that is puzzling and the question is "Why does this happen?" then it is explanation they are after. If the interest is centered on the status of the theory itself and the phenomena are just hoops it must jump, then it is confirmation. There is, of course, nothing wrong with a combined notion of explanation-as-

71

confirmation, and this is exactly what we did in Chapter 3 by considering explanation as a truth-conducive virtue. The focus in this chapter, though, will be on the external virtue a theory displays by having true, observable consequences, whether or not they are explanatory. The concern is not so much *why* the phenomena happen as it is *that* they happen as predicted by the theory.

Start with an example, Einstein's general theory of relativity. In its early years, as a proposed rival to Newton's well-tested and phenomenally successful account of gravity, the general theory of relativity is appropriately thought of as a hypothesis, a provocative suggestion in want of testing. It is a testable theory insofar as it makes definitive predictions of observable phenomena. It describes gravity as so ubiquitous a force in the universe as to be in the fabric of space and time themselves. Anything moving through space and time is affected by gravity. All material objects, planets, bricks, people, will feel the pull of gravity and will move accordingly, but this is old news; Newton already said as much. The new idea is that even the passage of light, since it moves through space and time, will be affected by gravity. A consequence of the new theory is that light will be deflected if it passes near a massive object since the light will be attracted by the object's gravity. For example, light from a distant star will be very slightly bent if it passes near the sun. The light is bent if it passes near a planet or even a basketball, but of these three only the sun is massive enough to bend the starlight by an amount large enough to detect. With the sun providing the gravity, the bending of light is observable by noting the relative positions of stars when the sun is well out of the way, that is, when there is none of the putative distortion by gravitational bending, and compare this with the relative positions of the same stars when

Confirmation

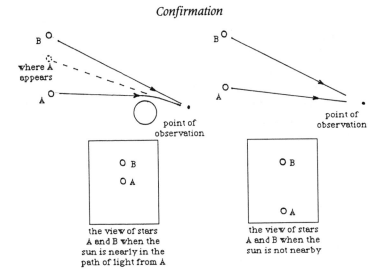

Figure 4.1. Two stars, A and B, viewed in two circumstances to see if the gravity of the sun affects the trajectory of light.

their light rays must pass near the sun (see Figure 4.1). Einstein's theory predicts that the sun, or any sufficiently massive object nearly in the path of light, will cause a distortion in the observed pattern of stars. The effect will be observable only during an eclipse of the sun, not only because no one wants to look right at the sun but also because the sunlight must be blocked for the nearby stars to be visible.

The first total eclipse of the sun to occur after Einstein introduced his theory, the first, that is, to occur during clear weather and without the distraction of a world war, was in 1919. The prediction about the observed positions of the stars was true. This was clearly good news for the theory, and it helped make Einstein everyone's favorite scientist. But just how good is the news? The prediction was confirmed, but was the theory? What's the difference? What does one true prediction prove? Suppose the prediction turned out to be wrong? Would that have been sufficient evidence to disqualify the theory once and finally?

Clues to answering these questions lie in the basic logical relation between the theory and test results. The initial model we are about to propose will need some fixing up to accommodate realistic examples, but it is a useful place to begin. The theory being tested, Einstein's theory in this example, is regarded in this case as the hypothesis, thus marking it as the claim in question, the one proposed with, as yet, no strong commitment to its being true. Testing it requires that observational consequences be deduced from the hypothesis, reasoning, as in the relativity case, that if the hypothesis is true then the world must behave in such-and-such a way and we should be able to observe this. This draws out the causal consequences of the theory and describes phenomena as determined by the theorized objects and events.

This scheme of finding the deductive consequences of the hypothesis in question is called the **hypothetico-deductive method of confirmation,** H-D for short. The deductive part is only the inference from the hypothesis to its observable consequence. If the hypothesis is true, then the effect must happen. That is, on the assumption that the hypothesis is true, what phenomena must follow? Observing the effect then, seeing that the predicted phenomenon does in fact happen, enhances the credibility of the theory, but it does not constitute unequivocal confirmation. It is not a deductive demonstration that the theory is true. It's not even close. To claim that it is would be to argue in the form that ''if the theory (the hypothesis) is true then the effect must happen,'' and ''the effect does, in fact, happen'' so ''the theory *must* be true.'' This is exactly the same fallacious form of reasoning encountered in analyzing the truth-conducive properties of explanation and the same fallacious argument used to reason that if it's summertime it must be the fourth of July.

The point is that observing a true consequence of a theory is not deductive, absolutely certain confirmation of the theory.

Hypothetico-deductive is not intended to mean a deductive proof, nor does it suggest the deducing of the theory from observational evidence. It is rather the other way around, the deducing of observational evidence, predictions, from the theory. True predictions then provide only inductive support for the theory by increasing the likelihood that it is true. If one successful H-D test helps a little, then further H-D tests will help more. The general theory of relativity does not rest its credibility on the single success of predicting the bending of starlight. It makes other predictions, for example, that gravity affects the passage of time and the frequency of light. This is the gravitational red shift, and it too has been observed. In clear hypothetical-deductive form it contributes inductive support to the theory.

This pattern of H-D confirmation is likely to be just an articulation of common sense. Everyone knows that one successful prediction does not prove beyond all doubt that a theory is true. But what about one failed prediction? Does that unequivocally disprove a theory? Suppose that observations during the 1919 eclipse had been contrary to Einstein's prediction.

Since we are still talking about a hypothesis and its deductive predictions, it makes sense, in the case where the prediction fails, to speak of hypothetico-deductive falsification. If the hypothesis is true the effect *must* occur, but the effect does not occur, so the hypothesis must be false. This is a valid deductive argument, and it does prove with certainty that the hypothesis is false. The idea is that if the theorized object or event is *always* supposed to cause the observable effect mentioned, but in fact the effect is missing, then the theorized object must not be there. There is no leeway in this argument, no degrees of unlikelihood. The case is closed. On this model of H-D falsification, a single failed prediction will suffice to falsify a theory and force its rejection.

There is, apparently, an asymmetry in this model of testing between confirmation and falsification. Confirmation comes in degrees and is vulnerable to confusion as to how much is enough.

Falsification, on the other hand, shows no such sloppiness, no degrees, no need for discretion. Once is enough. By virtue of the logical pattern of drawing predictions from theories, falsification has the advantage of a stricter standard, a deductive demonstration of failure. This advantage has motivated some who have come this way before, most notably Karl Popper, to emphasize falsification, that is, weeding out rather than cultivating, as the most crucial phase of the scientific process, and falsifiability as the essential feature of a scientific theory.

This basic model of H-D confirmation and falsification, though, is much too simple. It misses crucial details of real cases of scientific testing, and it is only because of this mistaken view of science that the asymmetry between confirming and falsifying appears. In real life, one failed test does not force the rejection of a theory, and the model of testing must be fixed to show how this works and to see if it is reasonable to retain a theory in the face of failure.

In its favor, though, the simple H-D model does capture the core of common sense about scientific testing. Examples like the general theory of relativity show that H-D gets at least the basics right. It reflects the reasoning of scientists in their plans for experiments to test theories, and many real examples of scientific testing clearly show the essentials of the H-D model. The relatively new suggestion, for example, that the electromagnetic force and the weak nuclear force, the cause of some forms of radioactivity, are in fact two manifestations of a single, electro-weak force needs to be tested. This new theory makes the novel prediction that protons decay. Any proton is likely to remain intact for a very long time, but it could eventually split spontaneously into other particles. The theory precisely describes the decay products and thus suggests observable consequences. To test the theory, look for the characteristic traces of proton decay. Thus from theory to experiment. It is a complicated experiment, requiring keeping tabs on lots of protons, since each one is very unlikely to decay. With so many protons and so few decays, the results are, so far, unclear. Experimenters are indeed following the H-D pattern of checking the consequences to test the theory, but observations of the consequences have not been conclusive.

Another good example of hypothetico-deductive testing is the search for solar neutrinos. The standard and well-entrenched theory of the anatomy of stars, including our sun, has it that nuclear fusion is the source of stellar and solar energy. Nuclear fusion is well understood, at least to the point of knowing that the process produces neutrinos which, according to theory, should easily escape from the solar interior where they are born and some of them, a predictable number, should find their way to earth. Furthermore, enough is known about neutrinos to know how to detect them. It isn't easy, but it's theoretically possible. It is done with a huge tank of cleaning fluid, deep underground to provide shielding from other distracting particles. Neutrinos occasionally interact with the chlorine in the cleaning fluid, changing it into a radioactive argon atom. The radioactive emissions from the argon can then be counted as the footprints of solar neutrinos.

This is fine H-D form. If the hypothesis about solar fusion is true, there must be detectable solar neutrinos. So test the theory by looking for the neutrinos. The results in this case are not at all ambiguous. The experiments to detect solar neutrinos have been on the air for years, and yet they have not detected any neutrinos, or at least not nearly as many as the theory predicts. This is a frustrating result that is well known to astronomers and physicists as the solar-neutrino problem, but apparently it is not a big enough problem to force the rejection of the theory, and contrary to the naive model of falsification, the failure to observe the predicted effect has not been taken as conclusive proof that the theory is false. Furthermore, this is not an unusual case of testing in science. There is always something to blame, perhaps in the experimental setup or in the reasoning process that issued the prediction. It is not, after all, a simple, two-component deduction from the hypothesis about solar fusion to the prediction of clicks on the detector counting radioactive argon. It is a more complicated argument than the simple H-D model indicates, and a realistic model of scientific testing must take the details seriously.

To judge by the complication and expense of scientific experiments, the predictions being checked are not simple statements of the form "*x* happens." The prediction usually specifies how to set up the experiment and how to detect the effect as caused by the theorized events. It specifies the proper conditions under which the testing and viewing are to be done. The prediction, in other words, is a conditional statement. *If* things are set up in this way, and *if* the observing conditions are right, *then* we will see *x*. Recall the balloon theory, the one that explains why rubbed balloons stick to walls by the scraping off of old, crusty rubber, exposing fresh, sticky rubber. To test this theory one might use the theory itself to predict that a rubbed balloon will crust over again if left to dry and so the balloon will no longer stick after a sufficient drying time. The prediction is a conditional, an if–then statement. If the balloon is left alone for a while, then it will no longer stick to the wall. Both the specified condition, leaving the balloon alone, and the result, its not sticking, are events that can be observed.

As with the solar-neutrino experiment, the prediction being checked is complicated. In terms of the objects and events amenable to observation, it says something like this: If you put a tank of cleaning fluid underground, and if you periodically flush it and separate out the argon, and if you count the radioactive decays, then there will be *x* number of counts. There are a lot of conditions that must be properly satisfied in order to bring about the final result of meaningful clicks on the counter.

If the results do not happen as expected, if there are too few clicks or if the balloon still sticks to the wall, we *can* blame the conditions. Maybe the experiment was set up incorrectly. Maybe the viewing conditions were wrong, or there was some unnoticed interference with the experiment. In the case of the balloon, if it still sticks to the wall, there might still be truth in the theory that rubbing scrapes off old rubber and exposes a fresh, sticky surface, but in the testing process the balloon wasn't left long enough to dry, or the air was too cold and drying was retarded. These could be responsible for the expected result failing to appear, and until these issues are settled, we are not forced to reject the theory.

The form of the deduction in this revised hypothetico-deductive model now has three components, the hypothesis (that is, the theory being tested), the conditions of testing, and the predicted effect: If the hypothesis is true, then if conditions are right the effect will be observed. If in fact the effect is not observed, then we are forced to conclude that *either* the hypothesis is false *or* the conditions are not right. Blame the conditions if you want to retain the hypothesis. If there are too few neutrino-indicative clicks, maybe the counter or the argon-separation process is inefficient. The fluid may be contaminated or perhaps the tank is not deep enough or it is too deep in the earth. Lots of things can go wrong with experiments, and failed predictions do not automatically indicate a false theory.

The effect might be there, in other words, but we are just not seeing it because we are not doing the experiment properly. This is not an unreasonable excuse, given the complication and sensitivity of most scientific experiments. It's not so unreasonable in simple cases either. Theories of gravity predict that if two objects of unequal weight are dropped simultaneously from the same height, they will hit the ground at the same time. But if you try the experiment it turns out that they don't hit at the same time. The problem is not with the theory. The problem is that the experimental condition (there must be no air resistance) is not satisfied.

As long as the experimental conditions are observable, however, they can be checked, and blaming conditions is not unresolvable speculation. It is rather a suggestion to have a look and see, and to do the experiment again with the conditions right. Drop the two things in a vacuum. Check the efficiency of the counters. That is the business of the experimentalist, to keep tabs on the conditions and to specify standards by which to tell when conditions are right. So there are limits to how far you can get by blaming the experimental conditions. This limitation, though, will be less effective on the next detail that must be added to the H-D model of testing.

If we think back on the example, it is clear that more than one theoretical claim is needed to deduce the predictions. The hypothesis alone does not entail the prediction; it is rather the hypothesis in conjunction with **auxiliary theories** from which predictions are deduced. The realistic model of testing begins with the deduction that if the hypothesis is true *and* other theories that describe mechanisms relevant to the manifestation of the effect are true, then if conditions are right the effect will be observed. Of all these theories used to deduce the prediction, there will be one theory distinguished as the hypothesis, that is, the theory we are interested in testing, but it is important to keep in mind that this is a distinction that depends on the interests of the scientists. All the theories used in the deduction, hypothesis and auxiliaries alike, are about unobservables, and in this sense all are taking risks and are at some point in need of confirmation.

Consider the case of solar neutrinos where the hypothesis is the claim that nuclear fusion is at work in the solar interior. This claim by itself does not entail any predictions about neutrinos or detector clicks. The hypothesis doesn't even mention neutrinos.

The deduction of the crucial prediction can be done only with additional premises. There is a premise to the effect that fusion creates neutrinos, and this claim is a theory, one of the auxiliary theories needed for the deduction. Furthermore, they are not just any old neutrinos but specifically electron neutrinos (there are two other kinds) produced in the sun. Once produced, the neutrinos are sufficiently elusive to escape from the solar depths where they are created and to speed to earth. This is another theory required for the deduction. There must also be a theoretical account of how the neutrinos interact in the tank of cleaning fluid, how the argon is chemically separated, and so on. All of this theoretical information is needed to draw the experimental consequence from the solar theory. That's why science isn't so easy that you and I can settle in an instant the issue of nuclear fusion in the sun by simply looking for neutrinos in our cleaning fluid. You have to know a lot of theory in order to do meaningful experiments.

The logical structure of the deduction, then, is that if *all* of these theories are true, the theory of solar fusion and of fusion producing neutrinos and of the great mobility of neutrinos and so on, then the effect must happen. If all goes well and the predicted effect does happen, this contributes to the inductive evidence in favor of the hypothesis, the claim about the source of solar energy. This complicated case with auxiliaries and experimental conditions is not a new case. It is a more faithful representation of every case, an improvement on the original, basic model of H-D confirmation. A positive test is not unambiguous confirmation of the hypothesis, but it helps.

What if the effect doesn't happen? Suppose, as is in fact the case, there aren't nearly enough neutrinos observed and yet all the prescribed experimental conditions check out as being fulfilled. The deductive argument of falsification forces the conclusion that, without a doubt, not all the theories, hypothesis plus auxiliaries, are true. That much, but only that much, is known for sure. It is unlikely that they are all false, and surely this extreme conclusion is not warranted by the evidence. But is it the hypothesis that is false or one of the auxiliaries? Don't forget that the auxiliaries, even if previously tested, have not been unambiguously confirmed. They are not above suspicion. The negative result of the experiment deductively proves that one or more of the theories is false, but it does not, indeed it cannot, indicate which one it is. There is no unambiguous falsification of the hypothesis. By blaming an auxiliary we can retain the hypothesis, and the possibility of its being true can be kept alive in the face of a false prediction.

This is exactly what is happening these days in the solar-neutrino problem. Virtually nobody wants to reject the theory about nuclear fusion as the energy source of stars. It is a firmly entrenched account that explains a lot about stars, the abundance of the elements in the universe, and the evolution of the universe. Rejecting this theory would force a great deal of rewriting in physics and astronomy. Some of the auxiliary theories, on the other hand, are not so securely entrenched, and their rejection would upset less in the web of scientific knowledge. For example, the claim that the neutrinos travel virtually uninhabited from the sun to the earth could be wrong. There could be some unseen matter, the speculated weakly interacting massive particles (yes, they are called WIMPS), interfering with the neutrinos before they can get here. Or the theoretical claim that the neutrinos escape unscathed from the center of the sun could be mistaken. Any subtle influence that changed them from electron neutrinos into either muon or tau neutrinos (and such changes are theoretically possible) would render the neutrinos invisible to the detectors presently used. These are the sorts of theoretical claims, the auxiliaries, that are drawing the doubt as a result of the solar-neutrino problem.

It is important to understand the difference between the auxiliary theories and the experimental conditions. Though either or both can absorb the blame for a falsified prediction, the conditions are directly checkable (and often correctable) whereas the auxiliaries are not. The inclusion of auxiliaries introduces a lingering ambiguity in cases of falsification, since the auxiliary theories are not directly checkable. The observations are never clear as to which of the theories, hypothesis or auxiliaries, is the culprit, and this reveals the importance of internal virtues. The decision about which of the theories to question first is often guided, as in the solar-neutrino case, by consideration of internal features such as entrenchment, explanatory coherence, and simplicity.

The realization that the testing of a theory against the observable evidence can yield neither certain confirmation nor unequivocal falsification indicates the impossibility of crucial experiments. A crucial experiment would be a test that takes place in the context of two rival theories that offer conflicting descriptions of the same aspect of the world. Light-as-a-wave or light-as-a-particle was once a question of competing theories. Earth-centered or sun-centered universe was another. And without stretching the idea too much, any theory (there are x's) and its denial (there aren't x's) are rival hypotheses. A crucial experiment would be a definitive test that would clearly rule out the possibility of one theory and thereby rule in the other. But such clarity never happens, and expecting it is the product of a naive view of the falsification process. Either hypothesis is retainable regardless of the experimental outcome. It may require rejecting or even adding some auxiliary claims, but it can be done.

This is not to say that we never decide or cannot decide between rival theories. It happens all the time. The point is simply that the observational evidence alone does not force the decision.

Here is a final but classic example of retaining a hypothesis in the face of a failed prediction. Not only were the auxiliaries doubted before the hypothesis, but a new, ad hoc auxiliary theory was devised to account for the unfavorable experimental outcome. The whole affair had a very profitable result.

In the nineteenth century, Newton's theory of gravity was tested through its predictions of astronomical events. The theory (or, more carefully, the hypothesis, since it was the theory being tested) predicted eclipses, the comings and goings of comets, and the orbits of the planets. In fine H-D form, the theory was put to the test by observations of the predicted positions of the planets, all seven of them. The theory made lots of true predictions and accumulated lots of inductive, H-D support. But then it made a mistake. The predicted orbital positions of the planet Uranus did not match the observed positions. But this false prediction did not force the rejection of Newton's theory, as our preliminary, naive account of falsification would have suggested.

To retain the theory, though, one must fix the blame for the false prediction either on the experimental conditions, saying the prediction isn't really false (things just looked that way), or on the auxiliaries, saying the prediction is false but the hypothesis (Newton's theory) is not to blame. Appealing to the experimental conditions was not a lasting option, as the measurements and observations were demonstrably accurate enough to show that Uranus really wasn't where it should be.

The auxiliaries then. What are they? To calculate the position of a planet one must know not only about the gravitational force that affects it but also that there are no other forces of consequence acting on it. It is also necessary to know the sizes and positions of the nearby massive objects causing the gravitational forces. In this case that means a model of the solar system, and it was this model that drew suspicion. Even if the gravitational force works exactly as the theory says, we will get erroneous predictions if we do the calculations with erroneous data on the objects causing the gravity. Errors of this kind would result if the mass of Uranus is misunderstood, or if we are wrong about its orbital radius. Or maybe there is yet an eighth, as yet unseen, planet beyond Uranus, which is pulling it from its predicted orbit. In fact, the position of this putative planet, where it would have to be to cause the deviation, was calculated, and in 1846 Neptune was discovered, right where it should be.

The decision to stand by Newton's theory of gravitation and to question the auxiliary theories instead was guided by internal factors. Newton's theory was richly explanatory and elegantly efficient. It made good sense of a lot of what goes on in the universe, and rejecting it would force a great deal of rethinking and reworking of the understanding of the physical world. The model of the solar system, on the other hand, was not so crucially connected to the theoretical understanding of the universe. It could be rather easily revised, and so it was. The decision to question this particular auxiliary theory rather than the hypothesis in turn influenced the selection of further tests; thus internal factors direct not only the justification of theories but the choice of experiments as well.

SUMMARY

The moral of the story is threefold.

Theories are compared to the evidence in groups, never one at a time. In the most realistic model of scientific testing, the hypothesis and the auxiliaries are equally on the line. They are equally vulnerable to the data. As a result of this, science admits of no unambiguous falsification just as it has no facility for unequivocal, dead-certain confirmation of theories. It is always the consequences of a theory that are compared to the evidence, and between theory and its consequences there is always room to equivocate.

The second important point in the analysis of confirmation has already slipped out. In science, there is no conclusive proof or disproof of a theory. Scientific proof is not conclusive proof. This is not to say that there is no justification of scientific theories.

85

Indeed, the likelihood of a theory being true is influenced by the positive or negative outcome of a test, as well as by the evaluation of its internal virtues. But the loss of certainty should be a warning to the superscrupulous who expect nothing less than certainty from science. If we hold out for certainty, we get nothing, and we will miss the crucial activity of science that is the evaluation of degrees of likelihood of theories and the warrant for believing that they are true. It is a comparative analysis, assessing not absolute justification of theories but a relative scale by which to judge among competing claims.

Finally, it has become clear that science must rely on more than just the observable evidence to adjudicate the warrant for theories. Science is much more than simply letting the facts decide. Internal factors are an essential and influential part of the justification process. They must be, since the observable data do not contain enough information to determine which theories are true. The observable evidence, as it is sometimes said, underdetermines the theoretical account of the world.

5

UNDERDETERMINATION

THIS is where Chapters 3 and 4 have been leading. In the terminology of internal and external virtues, external virtues do not determine that a theory is true or that it is false. A theory can be explanatory and can be favorably tested against the observable evidence yet still be false. The logical structure of explanation and testing is never of the form that the evidence entails a theory. Rather, it is the other way around. A theory entails its evidence, and because of this, observation of the evidence cannot stand as clear confirmation of the theory. Nor can the evidence provide unambiguous falsification since, as we have seen, realistic cases of theory testing are of the form, theory *plus auxiliaries* entail the evidence, thus confusing the issue of assigning guilt in the case where the predicted event does not occur.

To put it bluntly, the observations of the world, that is, the reports on the manifest phenomena, are not sufficiently informative to single out the one true theory of what is going on behind the appearances. This is not to say that anything goes, that one theory is as good as another, or that observations have no significant role to play in the scientific justification process. Explanation and testing are strict and crucial discriminators of scientific theories in that a theory is unacceptable if it is in no way explanatory and does not cooperate with other theories to make true predictions. The comparison of a theory with observations weeds out lots of potential theories. The problem of underdetermination is that this process doesn't weed out enough, since many competing theories can account equally well for the evidence. The evidence alone cannot trim the pool of applicants down to the single theory that is the true theory.

A theory with a perfect score on the evidence, in other words, is not necessarily true. It could explain all relevant phenomena, leaving no outstanding puzzles, and it could be confirmed by all tests and still be false. There will always be an alternative theory that scores just as well on the evidence, an alternative that is empirically equivalent. This other theory explains the same phenomena and makes the same predictions but does so with a different description of what is happening in the unobservable world. At most, one of these alternatives is right about how the world works, and insofar as we are interested in this basic understanding, we have a stake in determining which one it is.

It is not necessary for us to be aware of the alternative theory or for scientists actually to use it to generate predictions for it to upset the process of justification. There are plenty of situations in science where only one theory is imaginable. But this is just a limitation on imagination, and as long as an alternative account is conceptually possible (and it always is), the observational data alone can give the theory we have in mind at best a fifty-fifty chance of being true. In fact, though, the problem of underdetermination is much worse than this because, for any finite amount of data (things explained, tests passed), there will always be *any number* of theories that are empirically equivalent over that data. Of course, the huge majority of those alternative theories are really dumb. Shy, moonstruck sirens rising from the oceans to cause the tides? Old, crusty rubber being scraped from the balloon to make it sticky? There is no doubt that these are silly theories, and no one in his right mind advocates them as being true. I don't and neither should you, but why? They are dumb theories as judged by internal standards. For example, they are far-fetched and wildly implausible, which is to say they violate all standards of entrenchment. The theory of sirens at least is unnecessarily baroque in its ornamental descriptions of humanlike creatures with emotional lives and intentional activities. Plain old gravity will do just as well. Plain old gravity, that is, is simpler in its explanation of the same phenomenon. So it's not the evidence that is decisive between these rival theories. It can't be; the crazy alternatives are empirically indistinguishable from the real theories (or if they aren't, then we'll find some that are).

If there is any number of empirically equivalent alternatives, then the observational evidence alone cannot give any particular theory even a fifty-fifty chance of being true. The pool of empirically adequate theories is as large as you like, and it can be trimmed toward the truth only in cooperation with internal standards. It is crucial, therefore, to be sure of the truth-conduciveness of the internal standards.

As a graphic example of underdetermination, recall the simple experiment of Chapter 2 in which the length of a metal bar was measured as the bar was heated. This example is far down at the simple end of the spectrum of complexity of scientific activities, and so it is not fairly characteristic of the complication of experiments or of the general difficulty in discovering alternative, empirically equivalent theories. It is, nonetheless, generally descriptive of the information imbalance in every case where theories meet the world. There is a finite amount of data, and this is always the case. All the past observations of things to explain, plus all the future tests to be made, sum to a large but finite number of data points. The number of theories that can successfully explain what wants explaining and that pass the tests, on the other hand, is unlimited. Among these empirically indistinguishable theories will be a variety of alternative descriptions of how the unobservable world works.

The thermal-expansion example is such a good representation of the imbalance, not because it is a particularly underdetermined experiment but because it is particularly amenable to graphic presentation and because the key moves in the explanation and confirmation activities are clearly identifiable. Figure

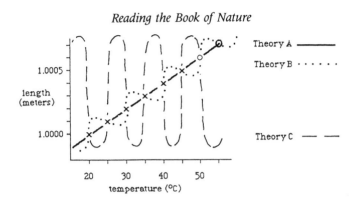

Figure 5.1. The heated-bar example (a reproduction of Figure 2.2).

5.1 repeats the graph of the data given in Chapter 2. The past data, things that have already been measured, are displayed on the graph as crosses, and the future data, outcomes of future tests, are shown as circles. A theory of what is going on here, that is, of the mechanism by which heating the bar causes it to change size, will describe what length the bar *would* take at any particular temperature. The theory is descriptive of hypothetical cases as well as actual cases. It goes beyond the actual evidence; that's what makes it interesting and useful. In this case a theory will provide a continuous line on the graph.

The theory represented by line *A* explains all the observed events and correctly predicts the future test at 50 degrees. Graphically, *A* passes through all the data points. But then, so does *B*. It is easy to see that for any finite number of data points there will be plenty of curves to choose from that pass through all the points. There will be an unlimited supply of empirically indistinguishable theories, but this does not mean that the scientist should advocate anything but theory *A*. It is foolish to conjure up curves in the account where none are needed. Theory *A* is preferable on standards of simplicity, and this shows off the importance of the internal feature of the theories.

90

The experiments of measuring the length and temperature of the bar demonstrate that the world *seems* as if theory *A* is true. There is no doubt that things appear this way. But, based on the observations, it appears as well as if *B* is true, and clearly both cannot be right. This forces a close look at the question of the inference from evidence to theory, the inference from "seems as if" to "is." Even when no alternative theory *B* has been imagined, there is reason to doubt the move from the evidence that the world seems as if *A* to the claim that the world really is *A*.

What is the responsible thing to do in the face of this doubt? One solution is simply to avoid the inference altogether by precluding science from making claims as to how the unobservable world *is*. Science, from this perspective, is in the functional business of reporting what the world seems like. It still deals in theories about unobservables, claiming with conviction that, by all experimental accounts, events in the world behave *as if* there are atoms (just as the events each Sunday morning are *as if* my neighbor steals my newspaper). Scientists are committed to the truth of such "as-if" claims, but this does not force a commitment to the corresponding "is" claim. Science, by this view, does not claim that the brass bar really expands in a linear response to rise in temperature, only that the bar is *as if* this is the case, just as it is as if the oscillation theory *B* is the case. What counts in science is finding an empirically adequate theory, and *A*, *B*, or *C* will do just as well. Truth about the unobservable is beside the point. It doesn't matter which of the empirically indistinguishable theories *A*, *B*, . . . , is adopted because they are only valuable for empirical purposes of explaining and predicting. For these purposes the alternatives are equally adequate, and there is no pressing decision to be made in choosing one over the others.

This response to underdetermination, diminishing the aims and responsible accomplishments of science in proportion to the limited informational content of experience, comes in two basic varieties. There are interesting variations on each of the two approaches, but these two perspectives will sufficiently broaden our view of the issue. One school of thought has the attitude that science doesn't (and shouldn't) intend theories to be thought of as being true or false. The true-or-false property is not a feature of theories. Theories are useful or not, applicable or not, but they are not true or false. This view of theories as useful, intellectual tools is often called **instrumentalism**.

Alternatively, theories might be seen as genuinely true or false, but underdetermination may prevent our ever knowing which score a particular theory deserved, true or false. When scientists theorize about atoms they really mean to be talking about real things. They *intend* to be telling the truth, but in light of under-determination, the responsible thing to do is to avoid committing oneself one way or the other. The theory is true or false, but we cannot tell which. This is the position of the **empiricist**.

These two views of science and the status of theories are two varieties of antirealism. This is an umbrella organization defined by its opposition to **realism**, a view that the theories of science are in fact true or false of the world, or at least approximately so, and that we can sometimes tell of a theory whether it is true or false. In some cases, according to the realist, the inference from "seems as if" to "is" is justified. Realism clearly bears the burden of proof to show just how this inference is to work and to specify for what kinds of cases the inference is justifiable. Realism has the responsibility to deal with underdetermination. To see how this burden might be carried, we should look closely at the two versions of antirealism and the most persuasive realist response to each.

THE RESPONSIBLE AIMS OF SCIENCE

The key to instrumentalism is the focus on the proper intent behind the use of a scientific theory and on the appropriate question to ask in evaluating theories. The proper question to ask about a

Figure 5.2. A model of the diatomic molecule *AB*, with the binding force represented as a spring.

theory is not whether it is true or false, because theories simply do not have this property, just as a hammer doesn't have the property of being true or false. A hammer, like any other tool, is judged for its usefulness and its applicability to a particular task. A hammer is more useful than a potato, for example, for driving nails.

Theories too are tools, by the instrumentalist account. They are intellectual tools for purposes of prediction of future phenomena and organization of observations. This is their intended role in science, and so the relevant criteria of evaluation of scientific theories are criteria of service, of usefulness and applicability rather than truth. Empirical adequacy – efficient organization and successful prediction of phenomena – is the goal of theorizing. There is no intent, indeed, no reason, in the mind of the instrumentalist, to push theories beyond their means by attempting to use empirical adequacy as evidence for truth. Equally empirically adequate theories are equally acceptable. As there is no one true hammer, there is no one true theory of gravity.

On this view, theories function as pedagogical tools for experts. Teachers of science will often invent intellectual devices to help their students understand the physical world. It is not uncommon in an elementary chemistry class, for example, to be given license to picture the atoms in a molecule as being connected by inner springs (Figure 5.2). This is a harmless aid to understanding the basics of physical chemistry. If you think of there being tiny springs between the atoms, you'll get the calculations and

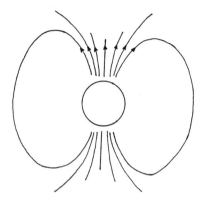

Figure 5.3. Magnetic field lines of the earth's magnetic field.

predictions right, and a lot of what molecules do, things like their oscillations, specific heat, and optical properties, will make sense. It is a very useful way to picture the molecules, but no chemistry teacher means to say that there really *are* tiny springs between the atoms. The claim about the real world is only that it is *as if* there are such springs.

Experts, not just students, use these sorts of conceptual aids. The magnetic field of an object, the earth for example, is often pictured and thought of in terms of the magnetic field lines (Figure 5.3). These lines are incredibly useful. Rotatable magnets like the needle of a compass will line up tangent to a field line. The density of lines, the number per volume, is related to the strength of the magnetic field, and an electric current is produced in a loop of wire if the number of lines that pass through the loop changes over time. Physicists often speak of "cutting field lines" to describe passing the wire loop in and out of the magnetic field. But it would be wrong to think that just because you can cut them they are real. No one even suggests that there really are lines around a magnet or around the earth. The references to magnetic

94

field lines are not intended to be taken literally, and so it would be inappropriate to call them true or to call them false. Claims about the field lines are simply useful. The description of field lines is an effective intellectual tool for dealing with magnetic phenomena, neither more nor less.

This interpretation of the status of scientific theories applies generally to all theories, according to the instrumentalist. Science has demonstrated very nicely that the world is as if there are atoms, tectonic plates, genes, a Minoan civilization, and so on, and that's enough. That is the proper goal of science, and with empirical adequacy the job is done. There is no question of whether the theories are true or false, of whether atoms really exist, because the theories are not meant to be true or false. Quarks are just like inner springs, only for grown-ups.

This view of the proper aims of science, the disavowal of any claim to a true account of the unobservable, is motivated by a desire to distance science from the occult. If appeals to mystical, unseen forces as in astrology or psychokinesis are to be disregarded as meaningless mumbo-jumbo, then one has to wonder how science is any different (that is, any better) if it invokes mystical, unseen entities such as electrons and genes. Instead of answering the question, the instrumentalist disarms it. The worry is dissolved if science simply stops talking about such things, or at least stops endorsing talk about unobservables like electrons and genes as true.

The instrumentalist move buys safety for the scientific enterprise, but it pays with interest. Instrumentalist science is no longer interesting science. It gives up what is most worth doing in science, namely, understanding what is happening behind the scenes in the realm of unobservables. The world, both those features that are manifest and those removed from experience, is in fact one way or another. Something causes the world as we see it. Observable objects are constituted of something. The past was one way or another. And in all these cases we want to know about it. This kind of knowledge, contrary to instrumentalism, is and should be the goal of theorizing.

Instrumentalism, in other words, asks too much of us in that it asks us to ignore our curiosity about the unobservable. More is expected of science than simply to drift in the shallows of empirical adequacy.

Of course, this tough talk is cheap. The imposing question is whether the interesting aspect of science, real descriptions of the unobservable world, can be recovered without undue loss of security. If theories are intended as more than pedagogical tools and calculating aids, if theories are supposed to be true or false, does science have the ability to distinguish the true from the false?

RESPONSIBLE ABILITIES OF SCIENCE

Grant science its aims of producing true accounts of both the observable phenomena and the unobservable causes and constituents. The cautious empiricist will do this but will also point out the problems of underdetermination. All confirmation, that is, all distinctions of the likely-to-be-true from the likely-to-be-false theories, will be seriously equivocal, and testing against the evidence cannot distinguish any one of the unlimited number of empirically equivalent theories. The empiricist will also point out the problems with explanation. Explanation in general, judged by its contribution to our understanding of things, is a pragmatic and psychological event with no clearly truth-conducive features. If we specialize to strictly causal explanation, the evidence still underdetermines which causal explanation is the right causal explanation. The extra kick from a causal account, recall, is what you get once you know that you are dealing with the *correct* causal explanation. To paraphrase Nancy Cartwright, if God tells us (that is, we are certain) that nuclear fusion *explains* the sun's energy, we can accept this and yet have no reason to believe that nuclear fusion is an actual solar process. But if God tells us that nuclear fusion *causes* the sun's energy, then we know as well that the nuclear fusion really takes place. This is good news as far as it goes, but of course God doesn't go around saying these things, nor does the natural evidence speak in such unequivocal terms of the correct causal explanation.

Given this underdetermination of theory by evidence, responsibility demands an attitude of agnosticism toward unobservables. Theories, according to a dominant group of empiricists, are indeed to be construed as true-or-false about the unobservable world, but it is most appropriate, by these empiricist standards, to withhold belief in the existence (or not) of unobservable entities and in the truth (or not) of theories about unobservables. The idea is not to say that there are no such things as quarks, for example. It is rather to say that we simply don't know and we in fact *can't* know if there are such things. The theory is in fact true or false, but we can't tell which. A claim about quarks or atoms, for example, is beyond the limits of justification and so beyond the limits of knowledge. Though there is good reason to use the theory *as if* it is true, there is no good reason to believe that it *is* true.

The description of this position as agnosticism is entirely appropriate because there is a significant similarity between questions of the existence of unobservable scientific entities and the question of the existence of God. We can ask whether talk of God is literal or metaphorical. Is reference to God merely a metaphorical way of talking about nature and the amazing (as well as the regular) things that happen in the world? Is talk of God just a way of dealing with otherwise overwhelming issues of life, death, and mortality? If we think *as if* there is a designer of our existence and a judge of our actions, then a lot of things make sense and there is purpose to living. Granting this, an atheist might claim that if there must be talk of God it should be construed as instrumental talk. There is no such thing as God, just as there are no springs attached to the atoms and no lines of magnetic force around the earth. If you must think in these terms to organize your affairs, that's fine. Just realize that it is only a useful invention you refer to as God, field line, or quark. In this sense, there are similarities between atheism and instrumentalism.

The more cautious agnostic would point out that, although God's existence cannot be proved, neither can God's nonexistence. There is no clear evidence that there is or that there is not such an entity. Therefore, either claim, of the atheist or of the believer, is unjustified, and in fact unjustifiable. The issue is beyond the limits of knowledge, and the appropriate attitude, the responsible attitude, is to withhold judgment. Thus the religious agnostic shares standards of justification with the scientific empiricist.

This parallel between reasoning in science and in religion is something to keep an eye on. God and electrons are both unobservable. Can we justify claims about one but not about the other? If we can prove that electrons exist, will the same style of proof demonstrate the existence of God? Why not?

The issue of scientific realism pivots on the distinction between empirical adequacy of a theory and the truth of a theory. Clearly these are distinct features. They mean different things. A realist view of science, though, sees the two as linked and must accept the burden to demonstrate this link. It's not that empirical adequacy (something that is directly testable) is simply the criterion of truth (a feature that is not directly testable), that is, that empirical adequacy is all there is to truth. The realist claims rather that empirical adequacy is a symptom of truth, meaning that compatibility with observations is an indication of a deep truth, truth about unobservables.

The standard presentation of the link between empirical adequacy and truth of a theory is known as the inference to the best explanation. Since it aims to exploit the truth-conducive virtues of explanation it is most effectively applied to causal explanation. Simply put, the realist claim is that there is good reason to believe in the truth of the theory that is the best causal explanation of the phenomena. Identifying the *best* explanation limits the collection of acceptable theories to just one, and insisting on a causal explanation insures that the theory is not explanatory only for its pragmatic qualities. The causal claim, recall, is definite in what it says is true of the unobservable world. An entity must actually exist to be a cause. It is not enough for it simply to *seem* to exist.

The realist position is that it is possible to justify some claims about unobservables by noting whether those claims function in the best explanation of observable phenomena. Being unobservable, in other words, does not automatically put something beyond the limits of knowledge. This is not an across-the-board endorsement of the belief in the unobservable entities of science. Some claims about unobservables are warranted and some are not. Furthermore, the justification for belief in a theory can change. Newtonian mechanics is a perfect example. For a long time the theory's explanatory and predictive success and its internal plausibility provided justification to believe the theory to be generally true. But both internal and external evaluations change, and the theory is now less likely to be true. It's not that its truth value changes. Either it is true or it's false for all time. What changes is our warrant to believe one way or the other. Under the circumstances, it was once justifiable to believe the theory. Circumstances change, and so there is no guarantee that what is justified now will always be. But you do the best you can; that's the responsibility of science.

The realist's cause, that inference to the best explanation is legitimately inference to the likely truth of the theory, hinges on the notion of being the *best* explanation. To license the inference, this bestness must be an evaluable feature and not one of these "if God tells you it's the best" arrangements. And there must be an explicit demonstration that being the best, in whatever sense defined, is a truth-conducive feature. Clearly, the standards of being the best will be largely, if not entirely, internal. Being an explanation is an external virtue since it involves comparison with observation, but being the best among empirically equivalent explanations will be a matter of being a simple, precise, and fruitful theory that is compatible and cooperative with other acceptable theories. Determining the best explanation will demand trading off some internal virtues for others and finding the optimum balance.

None but the most reckless realist would say that the best explanation is guaranteed to be true. The inference concludes only that the causal explanation that is most fruitful, that is, explains the most phenomena with the minimum of theoretical extravagance, and that fits into the larger theoretical network is more likely to be true than a theory that does none of this or does only some of it. It is unlikely that a false but adequate guess at an explanation of one phenomenon will both fit in with other scientific theories and contribute to explanations of other phenomena and other theories. The demand for coherence among theories and the goal of maximal coherence limit the number of possible theoretical pictures of the world and limit it in the direction of truth. The antirealist's sharp restriction of knowledge at the line of unobservability ignores this aspect of justification and preempts even comparative evaluation of, among theories of unobservables, which are more likely to be true.

Inference to the best (causal) explanation, in its most intuitive appeal, is this: The explanatory and predictive success of theories would be an absolute miracle if the theories were not true. It would be a miracle, which seems to happen over and over again in science, for a fictional account to fit consistently onto the reality of observation. This appeal has particular force in cases where several theories coincide, that is, where coherence among theories is clear. Consider the early experiments to measure Avogadro's number, the number of molecules of substance contained in one mole. This number is of course based on a molecular theory. The world is *as if* there are molecules, but does that indicate there really *are* molecules? In this case there is warrant to say yes. The number of molecules in a mole can be measured in a variety of ways, and predictions of the number can be generated from a variety of theories. It can be measured by chemical, thermodynamic, electrical, microscopic-motion (Brownian motion), and other techniques. All of these experiments come up with the same number. There is no collusion among these various theories; hence it would be a staggering miracle, a stroke of phenomenal luck, for them all to agree, unless they were measuring something real, namely, real molecules. There seems no accounting for this coherence among independently developed theories other than to give them a united basis in truth. If the molecules really exist and molecular theory speaks the truth, then agreement in counting molecules is to be expected.

This sort of reasoning is strongly analogous to methods used to translate unfamiliar languages for which no dictionary or lexicon is available. Scientists are in a predicament similar to breakers of codes or translators of ancient texts written in unfamiliar languages. On the basis of what appears, the printed words in the translator's case, the basic observations of phenomena in the scientist's, the task is to finesse the meaning of the story of what the world is like. An instrumentalist would claim that there simply is no single meaning of the book and one reading is as good as another, as long as it is consistent with the marks on the page, that is, with the observations. An empiricist of the sort we are following would say there is a single, real meaning. The plot is one way or another, but without meeting the author and without the definitive lexicon, we can never know what the meaning is. The realist, as characterized above, agrees that the book has a single, true meaning, and furthermore, the translator can hope to figure it out, just as science can figure out the single, true description of the way the world is. Some claims about the plot, some theories, will make more sense than others and will fit more easily and cooperatively with the claims about other aspects of the plot, and under the assumption that the book makes sense as a whole, only theories that fit a coherent account are likely to be true. Of course there are multiple coherent readings and hence multiple possible meanings, but as more of the book is examined and more pieces of the plot come to the light of theorized interpretation and must be fit with each other, insistence on coherence can shrink this collection in the direction of truth.

THE SIGNIFICANCE OF OBSERVABILITY

The differences between realism and antirealism are largely over the significance of the distinction between observable and unobservable events and entities. It is a question of the significance of this distinction as a limit of responsible knowledge claims. The antirealist, specifically the empiricist, regards observability as significant, since we are justified in claims about observables but not in claims about unobservables. We can judge as true or false claims about stones or clouds or planets but not claims about quarks or cosmic rays. There is warrant to believe statements about the Parthenon but not about Pericles. The distinction is between two distinct levels of justification. We are justified (according to the antirealist view of science) to believe claims about observables, but for things like atomic theory and other claims about unobservables, there is at best justification to use the claim as if it is true.

The realist view of science sees no sharp limiting of justification by the factor of observability. Justification can cross the line of observability. We want to know the truth about unobservables just as we want to understand the plot of the book and not simply know the shape of the marks on the page, and so there is need at least to compare theories and rate them as more or less likely to be true. This need is filled by the inference to the better or, in the context of more than two options, the best explanation. The realist will further undermine the significance of observability by

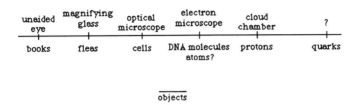

method of observation

unaided eye	magnifying glass	optical microscope	electron microscope	cloud chamber	?
books	fleas	cells	DNA molecules atoms?	protons	quarks

objects

Figure 5.4. A spectrum of indirectness of observation, getting more indirect to the right.

pointing out that the line between what is observable and what is not is itself unclear. If instruments such as microscopes and cloud chambers do not count as direct observations of microbes and protons, then what do we say of things seen with the aid of a simple magnifying glass or eyeglasses or even a windowpane? What about observing things through the air? The point is not that there is *no* difference between evidence viewed with the unaided eye and evidence dependent on the use of a microscope or a cloud chamber; it is rather that there is a difference of degree. Entities do not fall neatly into one of the two categories, observable or not; they occupy a more or less continuous spectrum of indirectness of observation (Figure 5.4). The empiricist has the obligation to explain where on this spectrum is the sharp division, or even hazy division, that separates what we can know about from what we cannot. Admittedly, there are clear cases of observables and unobservables. We can see stones, but we cannot see quarks. But so many of the interesting entities of science will not be so clearly either observable or not, and it is the justification of theories about these things that we are interested in.

The empiricist and the realist both have good points, and there is nothing to prevent us from taking what is good from each and synthesizing our own account. Clearly, the warrant for belief in a

103

theory generally decreases as the observational evidence is less direct. The more apparatus necessary and the more inference used to sanction the observational claim, the more chance of misrepresenting the world. The empiricist is correct in saying that observability is significant to justification. I certainly have better justification for believing in the existence of this pencil in my hand than I do for believing in the carbon atoms in the graphite that is rubbing off onto the page, and more still than for believing in the quarks that constitute the atomic nuclei. But the justification for theories does not abruptly fall away at any point along the spectrum of observability. There is no sharp limit of knowledge where the justification meter goes all of a sudden from full to empty. In other words, the realist is correct in that the significance of observability does not force a yes-or-no dichotomy of justification.

Benefiting from the good parts of both realism and antirealism, the most profitable approach to understanding scientific knowledge and scientific method is to look for *degrees* of justification. Measurement of these degrees will be in terms of internal virtues and external comparisons to evidence, and it will even be sensitive to degrees of observability of the objects and events referred to by a theory. Observation is a funny concept. We might have thought that either something is observed or it isn't and hence the entity is either observable or not. But there is more variety in the event of observation than that.

6

OBSERVATION

U NTIL the end of the previous chapter, we had been taking observation and the idea of observational evidence pretty much for granted. Some scientific questions, we have assumed, can be answered just by looking. Even if the logical implications about the truth of theory are unclear in the process of testing, the outcomes of the tests, whether the predictions hold true or not, have been taken as unproblematic. The test results, the observations, are of a purely manifest world of appearance. They are there for anyone who takes the time to look, and they can be used by anyone as impartial evidence. When in doubt about theory, there is always **observation**.

Observation is the touchstone of external virtues. Empirical adequacy, that is, compatibility with observation, is presumably easy and uncomplicated to evaluate. And observation is the distinction of external virtues, since it is the contact of theory to the outside world. Observation is the source of the world's input and guidance over our describing and modeling. We have separated external from internal virtues under a presumption that the world's input through observation puts the external features beyond our control in the sense that it's not possible to force the observations to come out one way or another. They are forced on us and we must take them as given. Observation is, if anything is, a source of objectivity in science.

This has been the expectation of the activity of observation that has motivated the internal–external distinction and has been quietly at work in the analysis of explanation and confirmation. It is an expectation that deserves a closer look, given the critical role observation is to play in the scientific process of justification. Objectivity, after all, is no insignificant burden. And in light of the complication suggested in Chapter 5, the notion of degrees of observability parallel to degrees of indirectness in the observation, the nature of scientific observation warrants a careful look. The concept of observation is used rather liberally by scientists, to judge by the sorts of things they claim to have observed. Protons have been observed, insofar as their distinguishing features, such as mass, electrical charge, and spin, have all been measured. And of the six varieties of quarks, it is not unusual to hear that five have been observed. The search continues for the one called the top quark. (Interestingly, it is also known as the truth quark. The search continues for the truth. Up, down, strangeness, charm, and beauty are all in, but truth is a continuing holdout.) The new Hubble telescope now in orbit around the earth is supposed to be so powerful and unbothered by atmospheric distortion that it should facilitate the observation of objects so distant, receiving light that they emitted so long ago, that we could see these objects as they were at very nearly the beginning of time. We can observe the big bang with this thing.

This may be how scientists invoke the notion of observation, but are these examples of information straight from nature, free of any conceptual embellishment, the undeniable, impartial, common ground of objectivity? And if these examples do not play that role, are there any scientific activities that do? Is the pure observation that has been taken for granted a real possibility?

THE IMPORTANCE OF OBSERVATION

There should be no doubt that the activity of observation is an essential aspect of the scientific process, but it is wise to emphasize this so that the importance of observation does not get lost in the following analysis. Asking questions of the activity and demanding an account of the source and handling of information should not be interpreted as an attack or a disguised criticism. Observation, like science in general, is scrutinized because it is so important, not because we expect to find flaws.

The sciences under account here are one and all empirical sciences. Whether they are natural sciences like physics, biology, and geology or social sciences like psychology and anthropology, the claims they make are contingent on what is going on in the world. Whether it is the world of people or of elements, organisms, and the environment, it is an actual world that is to be described by science. Contrast this with claims made by nonempirical enterprises such as mathematics or logic. These describe how some of the basic relations among things *must* be; the truth of these claims is not contingent on the way things in fact *happen* to be, and the truth of logical and mathematical propositions can be assessed without having to look out at what's happening in the world. It is a logical truth that outside at the moment it is either snowing or it is not snowing. I don't have to look outside to know that this is true. It *must* be true. This difference between necessary truths and claims whose truth is contingent on the way things happen to be shows up in the process of reading a book, as well as in science and daily life. It is the difference between descriptions of how books in general must be written, rules of grammar and noncontradiction, for example, and descriptions of how in fact *this* book that I am reading is written and what the writing says. The only way to tell what this particular book says is to read it, just as the only way to tell if it's snowing at the moment is to have a look.

Reading the book of nature, our metaphor for doing natural science, requires observing the events in nature. To have any claim to an accurate report of what's going on in *this* world, there must be input from the world. Observation, in other words, is a dietary staple of empirical science.

107

As the channel of information from the world, a great deal of expectation is placed on observation to be the decisive and authoritative arbiter of truth in science, and it is exactly this hopeful role that we are looking into. Empiricists in particular might hope that the reports of observation will serve as a common ground of agreement for all participants of science regardless of their theoretical or skeptical preconceptions. Observations, being imposition from the world, are intersubjective and public in a way that beliefs and theories don't have to be. They are accessible by everyone and, the hope continues, agreeable to everyone. As a common measuring stick, observations can be appealed to as adjudicators of theories to convince skeptics and those of alternative theoretical persuasion. When all else fails, in other words, and all manner of persuasion is inconclusive, we can at least all agree on the observations. They are the terminus of dispute.

This expectation of neutrality derives from a presumption that there will be some basic form of observation that is free of interpretive influence. The usual contrast is between observation and inference where the latter is the general description of the human contribution. When inferences and interpretations are involved there is the danger of error and bias, and the products of thought could be mere artifacts with no correlation to the real world. Observation, in contrast, is hoped to be just the facts, the unaltered information from the world.

There is the hope of the empiricist, and intuitively there is plenty of plausibility with the idea that there are some observational claims that any reasonable, unimpaired person will agree to. To strengthen the pull of this intuition, suppose we are given a questionnaire of true–false questions about items in the room. Half the questions are about the books in the room, stating such things as: (1) More than half the books in the room are blue. (2) None of the books in the room is on the floor. (3) There is at least one book in the room. And so on. The other half of the questions concern the cosmic rays in the room, and the statements to be judged true or false here are quite similar to those about books: (1) There is at least one cosmic ray in the room at any one moment. (2) On average there are more cosmic rays going down than there are going horizontally. (3) There are fewer cosmic rays in the room on Sunday than on any other day of the week. Making responsible decisions as to which claims are true and focusing on the features of a claim that indicate that it is likely to be true are exactly the tasks of justification. Science is like a true–false exam as we must responsibly decide, among proposed theories, which to endorse as true.

The telling factor in this questionnaire is not which of the statements are true or which ones can be answered. The important issue is which of the answers can be checked and agreed upon. With the books in the room, this checking and agreement come easily. Just have a look. Count the books, note their color, check the floor. Anybody can do this, and anyone who does and who is reasonably participating in a search for the truth will agree to the right answer. Books are observable. Cosmic rays are a different story. To check the answers to the true–false questions about cosmic rays we must either consult the relevant theories, those that describe what cosmic rays are, where they are likely to be, and what they are likely to be doing, or use a device such as a cloud chamber to detect the cosmic rays and reveal their activities. A properly prepared and oriented cloud chamber may reveal transient streaks of vapor, which are the telltale tracks of cosmic rays. As with the case of books, now all we have to do is have a look, point to the chamber, and say, "There they are! There are cosmic rays in the room." The similarity to observing books is shaken, though, with the question of why anyone should believe that these are in fact cosmic rays, in other words, when we question the responsibility behind the endorsement of claims about cosmic rays. Why do we agree with the report "There are cosmic rays," as we must if it is to function as a common ground of evidence? The answer is in theory, the theoretical account of the formation of the streaks, citing, as it does, cosmic rays as the efficient cause of the visible tracks. But the agreement, the common ground, disappears as long as there is an alternative theory of the formation of the lines. If someone suggests, for example, that the lines are formed by a magnetic effect and that what the cloud chamber shows are actually the magnetic field lines of the earth and nearby magnets, the claim about cosmic rays is not decidable without thought and the human contribution of interpretation. We are at the mercy of theory to check the answers about cosmic rays. But no one can pull any funny stuff about books. Whatever they claim about the books, we can just have a look for ourselves.

The empiricist's hope for observation, and the motivation for distinguishing observables (like the books and the streaks of vapor) from unobservables (like cosmic rays), is to escape from being at the mercy of theory. The hope is to identify some data that we can check for ourselves, information that is there for the taking because it is given by nature.

This tale of cosmic rays points out that the focus of our understanding of observation must be on the observational report rather than on the physical event of sensation. It is crucial to pay attention to what is claimed to be seen. The same physical event of looking at the cloud chamber could be reported as observing cosmic rays or as observing streaks. The difference in justification of the two reports is an important difference. We must attend to observational reports to see whether it is possible to fulfill the hope of escaping from the mercy of theory.

ACCOUNTABLE OBSERVATION

We are interested in events of observation that can participate in the process of science. To be useful as evidence an observation must make contact with a theory in some way if it is to serve as proof or as refutation. Science can make use of an observation only if it is an accountable observation, and this accountability has two aspects: The observation must be informative in the sense that it must be an account *of* something if it is to have the assertive content required to serve as justification of other claims. And the observation itself must be justified in the sense of being certifiably not haphazard or uncontrolled.

These two features of accountability are required of observations if they are to be informative and useful contributors to knowledge. To be evidence for a theory, an observation must be evidence *of* something. The proper question to think of observation as responding to is not simply "Do you see it?" but *"What* do you see?" or "Do you see that this is a that?" Accountable observation is more than just opening your eyes and saying "Yep, I see it." It requires a claim as to *what* is observed and an assertion of the *information* that is being conveyed by the world. One needs to know what it is evidence of if the results of observation are to be related to other ideas and compared to theoretical predictions.

Again, the lesson is to focus on the observational report. Even if the report is private, a report to oneself, the point is that the sensation must be described as a this or a that if it is to be useful. Simple observations such as the ones used to check answers to the questions about books in the room must be rendered in a conceptual description if they are to be compared to the concepts of the questions they are checking. A useful observation must be reported in an informational form, observing, for example, *that* the book is on the floor. Undescribed sensations are useless as evidence. Or perhaps it is better to say that pre-description sensations are pre-useful. Objects of thought require conceptual handles, and there is no way to think about ideas and to relate them to other ideas without the conceptual organization that comes with description. And if an observation report is to be relevant to a theory it must include some of the language of theory. It must assertively be evidence of an *x* if it is to be evidence in the case to justify a theory about *x*. To serve as justification, an observation must have some assertive content, and this is half of the requirement of accountability.

The other half is that the observation report must itself be justified if it is to have sufficient assertive authority to prove other claims. Science cannot accept just any old report of the senses. As a rule, observations must be carefully done and repeatable. They must be carried out in a controlled way and under the proper conditions. Furthermore, and this is important, one must be aware of the relevant conditions and the factors of care, repeatability, and control. It's not enough to insist on observations that *are* accurate or very nearly accurate. We must insist further that one *can tell* when they are accurate. We must insist, that is, on verifiable observations. We don't want to rely on dumb luck that the observation reports are accurate. Responsible observing is observing that we are able to justify. This requires an understanding of the relevant conditions under which an observation is done. It does not require that each individual observer understand these conditions, but the scientific community must understand what it takes to do the observation properly if the observation is to be acceptable as scientific evidence. Science, after all, is a public enterprise, and matters of justification are matters of community.

It is not that this verification report in fact accompanies every scientific observation, nor should it. The point is only that such verification is available. An acceptable observation report is one for which justification in terms of an account of the proper conditions and procedures *can* be given. In science, an observation cannot be entered as evidence unless it can be accredited in this way.

Thus are the two edges of accountability. To contribute to science or to knowledge in general, an observation must be about something. That is, it must be informative, assertive, propositional. And there must be some reason to think that the observational report is accurately about what it claims to be about. There is nothing absurd in the notion of an inaccurate observation, and so there is nothing out of line in the concern to block false observation reports from being entered as evidence. Beyond the immediately perceptual distortions such as dreams and hallucinations, there are observational distortions that result from unsuitable conditions. The observer may be improperly attentive or too far away. The viewing conditions may be too dark or littered with obstructions. Or there may be interference from outside sources causing a distorted view. It may all have been done with mirrors. In science, as in any responsible attempt to know about the world, we must be prepared to convince ourselves and others that none of these things has gone wrong and that what we claim to observe is an accurate report.

Recognizing the need for accountability of observation points out that observational information is not simply given to us by the world. Observation is not a matter of passive reception of ready-made, in the sense of ready-described and ready-justified, facts. Observation, insofar as it is a useful constituent of knowledge, is an activity of description and justification.

These are the general requirements of useful observation. The question now is, How are these requirements met in science?

OBSERVATION AND THEORY

The requirement that observation is to contribute to scientific knowledge, that is, that what is considered as an observation must be a knowledge event and not just a psychological event, has an important consequence: All observation in science is influenced by theory. This is not simply the way things happen to be done; it is the way things have to be.

Tying all observation to theory is a sweeping and a vague proposition, and it is of pivotal importance to the process of justification in science. It is therefore crucial to see the details of just how theory influences observation and how this affects the role of observation as objective adjudicator of the truth of theories. The suggestion is that what is observed in the world, that is, what is reported from observation, depends on the background theories held by the observer. Different theoretical backgrounds, as between someone advocating the caloric theory of heat and someone else believing the kinetic theory of heat, will lead to different observations in the pool of acceptable evidence. In other words, informative observation is impossible without a theoretical background, and alternative theoretical backgrounds lead to alternative informational reports. To paraphrase Kant, observation without theory is blind.

Enough with the slogans. It's time for details.

Theory influences observation in several ways. For one, our theoretical understanding of the world guides our selection of what observations to do. We cannot look at everything. Scientists must be selective of a few observations worth their time and money. The business of gathering data is neither comprehensive nor random, so it must be directed by theories that tell us where to look and what is important enough to warrant attention. Einstein's theory indicated that an observation of stars during an eclipse would be informative. Solar theory directs the observational efforts toward a search for neutrinos. This is of course the way things have to be and ought to be if the idea is to test a theory by its consequences. The point is simply to make plain the situation that the theories in place direct the selection of evidence. Different theories would precipitate a different collection of evidence. So when we talk about comparing a theory to the observational information from the world, the situation is more accurately described as comparison to a very small sample of information from the world, a theoretically specified sample. The sampling is done with an attention to relevant factors. That is, irrelevant variations need not be included. If we drop a brick and a sponge to test the parity of gravitational acceleration, we don't have to run the test with a sponge bought at Woolworth's and again with a sponge from Sears, or once with a blue one and again with a pink one. Where you buy the objects, certainly, and their color, probably, have no relevance to their gravitational or inertial properties, so these observations need not be made. Their exclusion from the collection of evidence is based on an understanding of the causal mechanisms involved, and that is a theoretical understanding. The relevance of an observation, and hence what is chosen to be observed, is influenced by the theories we believe.

A second way that theories influence observation is by guiding decisions as to when observations are credible. Theories play a key role in the accountability of observations by supplying standards to evaluate the reliability of observational reports. The assessment of the viewing conditions, attesting that there are no distorting factors or correcting for any distortions that persist, is based largely on an understanding of the causal mechanisms of observation. Clearly, there must be adequate lighting if a visual observation is to be credible. Why? Because vision is accomplished by receiving the light energy that bounces off the specimen and is transmitted through the space to the observer and into the eye. Light carries the information. Furthermore, it is no threat to the accuracy of the observation if the light must travel through air before reaching the observer. A theoretical understanding of the optical properties of air assures that, unless there is significantly uneven heating, light will pass straight through. And of course a well-made windowpane is no distortion to observation. It's not that each of us has to know a lot of theoretical optics to be able to look outside or to claim that it's snowing. It is rather that, if we are questioned on the legitimacy of the observation, there are these sorts of theoretical accounts to accredit the accuracy of the report. This sort of accountability is required of scientific observation, and it is clear that a different understanding of the causal interaction, that is, a different theory of what's going on in the event of observation, may sanction different observation reports as acceptable.

This role of theory is especially apparent in cases of observations that are assisted by machines. This is a very large class of important scientific observations. Telescopes, for example, if properly constructed, do not manufacture false images. They faithfully reproduce what is out there beyond the visual reach of the eye. In the case of observing the stars as a way to test the general theory of relativity, it wasn't the telescope that altered the apparent positions of the stars; it was a real effect of the sun. The telescope only magnified the view. Any distortions or alterations of the specimen can be accounted for and even corrected because we know how the machine works. The light from the stars is bent and focused by mirrors and lenses to produce an image with all of the relevant features of reality. This trust of accuracy is sanctioned by a theory of optics. Similar theories are the basis of our belief in the accuracy of images produced by a microscope. Reports about the microscopic specimen are not acceptable without a theoretical account of how the imaging is done. Such reports are not required with every reported observation, but they must be available. They are the answer to the question, Why should we trust the images produced by this thing? – should that question ever come up.

One doesn't have to understand the theory of how an instrument such as a microscope or a cloud chamber works in order to use the instrument. Noticing amazing things like the teeming life in a drop of pond water does not require an awareness of optical theory. You or I or anyone can use a microscope to see the single-celled creatures. Is this observation without theory? Biologists use microscopes routinely without really knowing how they work. And relatively unskilled and theoretically illiterate lab technicians are often told what to look for in the apparatus, and they succeed in observing what they should even with no theory in mind to influence their actions. In the good old days, the many photographs from particle detectors were scanned by hourly workers without a glimmer of understanding of particle physics. They were trained only to notice certain branching patterns in the streaks on the photograph and to note their frequency. Is this observation without theoretical influence?

Certainly not. A great amount of theory goes into the building of the machines, the particle detectors and microscopes, and into sanctioning their reliability. The data from these machines are acceptable, that is, the observations are accountable, only if the experts, those who invent and build the machines, have a solid theoretical account of how they work. Machine-aided observation is a communal event. Anyone can notice interesting features of the image, but the process is accountable only in light of the theories that describe how the image is formed. It is this accountability that gives the observation its license to contribute to knowledge.

It is useful to refer to these theories that describe the chain of interaction from the specimen to the observer as the **accounting theories**. They are not special kinds of theories with a special content. They are regular old theories with a particular use. Any theory could be used as an accounting theory. In cases of unaided vision, for example, basic theories of optics play the role of accounting. They answer for the accountability of the observation. With observing apparatus, the interaction chain is longer, as is the list of accounting theories. More theoretical claims will usually be invoked to verify that the image does indeed have information of the specimen.

The accounting theories, in other words, help clarify what the observations mean, in the sense that they specify what the images are images of. Accounting theories guide the appropriate description of the observation. Looking at the cloud chamber, for example, one is licensed to report that we see more than just streaks; we see images of cosmic rays, and the license is issued under the authority of a theoretical account of the cosmic rays, being electrically charged particles, ionizing the gas in the chamber, and thereby causing visible condensation along their flight path. The accounting theories fit the manifest image, the streaks in the cloud chamber, into the descriptive language of theory. This is a necessary step in the process of testing. Cosmic-ray theory will generate predictions about cosmic rays, and to test the theory these predictions must be imaged. Somewhere in the description of the observation, the test results must be about the objects of theory. This is how the observations are relevant to theory and can function as confirmation or refutation. Theory, in other words, guides the description of observation, lest the observation be irrelevant.

Even in cases where no apparatus assists the observation, describing what is observed has a significant theoretical influence. Simply calling something a this or a that requires having a concept of this or that. This application of concepts is done by grouping the individual items we encounter under headings of the same *kind* of thing. Things with relevant features in common are the same kind of thing. The individuals will not be exactly alike, but some features are irrelevant to membership in the category. Not all trees are green, nor do they all grow in Vermont, but these properties are not relevant to what it is to be a tree. Not all electrons are bound in atoms. There is a concept of being an electron that describes what it is to be that kind of thing, and that includes some individuals bound in atoms and some that are free. Specifying which features are relevant to the association of the kind of thing is a job for theory. Information on the grouping of items into kinds is not given by nature, since there are various ways to do it depending on which features are identified as important and relevant. This ruling on relevance is clearly influenced by theory, and in this way any activity of description, the application of a particular item to its conceptual kind, is guided by theory.

118

To summarize this final manner of theoretical influence on observation, all useful data must be described data, and the link between sensation and descriptive concepts is a link made fast by theory.

THEORY-LADEN OBSERVATION

"Theory-laden observation" is the slogan often used to summarize the points made in the previous section. Observations in science are theory-laden. Accountable observation, the only kind that plays a contributing role in the acquisition and justification of knowledge about the world, bears an inseparable theoretical component. We are forced to give up the notion of pure, unaltered information from the world. Better to think of observational information not simply as given by nature but as actively taken by the observer in an activity that is guided by the theories, beliefs, and concepts in the mind of the actor. This dashes the empiricist's hope of a common ground of evidence agreeable to everyone regardless of his or her theoretical predisposition. Different theoretical positions, perhaps rival theories that we are trying to decide between, could claim different evidential ground by authorizing different observations as important, different observations as reliably done, or even the same sensations differently described.

Accountable observation depends on accounting theories, whether or not the observation is done with the assistance of instrumentation. An observation's authority to justify other claims, that is, its usefulness, is based on its assertive content and its accountability. The spontaneous, uncomplicated result of just looking at the world may be that we come truly to believe what we see. But that's not good enough. To function as evidence in science, the act of observation must offer reason to think that we believe truly what we see.

119

The indelible theoretical component of observation means that observational beliefs are not rigidly foundational in the structure of justification. They are not immune to change or refutation. Observations carry relatively more authority than do more deeply theoretical claims in the sense that we are much more reluctant to deny the information of our eyes than information from more theoretical claims of things well removed from perception. As a rule, theories must bend to the mold of observation. But in the case of a single observation that conflicts with a large body of well-entrenched theory, it is an option wisely taken to discount the observation rather than theory. Better look again before chucking a healthy system of theories on the word of a single observation. It won't do to simply ignore the troublesome observation as if it never happened. Better to explain it, if possible, as a case of inappropriate conditions or, as a last resort, the distorting influence of some unknown interference. The point is that observations do not enjoy an absolute authority over theories. They too are acceptable insofar as they fit into the theoretical network. They are revisable under the influence of theories.

The picture then is not of observations dictating the acceptability of scientific claims. The relation between theory and observation is more of a republic where observations serve with authority by the will of the theories. As a populace chooses its rulers, theory vests observation with an authority to rule on the justification of theories. Observations in turn force theories to bend to their evidential rules, but they can force and bend only so far lest they lose their authority. Coherence among the theories, like consensus among the people, is still the ultimate source of persuasion.

The ambiguity inherent in the words "only so far" is exactly the same ambiguity in the process of falsification. In the event of an observation that conflicted with theory, a false prediction, one has to decide which theory, the hypothesis or an auxiliary, to reject. This is not a capricious decision. It is done with an eye toward maximizing the overall virtues of a whole network of theoretical claims. The current analysis of observation reports shows that they too are on the ballot for rejection or revision. One *could* reject the observation report. Again, the decision is guided by the participating theories and is done with a sensitivity to the plausibility of the accounting theories as well as the magnitude of disruption that would result if the observation is believed.

OBSERVING AND READING

The analogy between reading a text and doing science, used at the beginning of this chapter to accent the importance of observation, can serve as well to clarify the relation between observation and theory. In reading a book, we want to know the plot, the meaning of the text. The accomplishment of reading is to learn what is beyond the manifest marks on the page, that is, to understand what the marks indicate, what they mean. This of course cannot be done without carefully looking at the marks. So it is in science. The scientific accomplishment is to understand what the observations indicate, and this is done only through a careful attention to the observations themselves.

Marks on a page are not helpful at all unless they can be organized as letters, words, and whole statements. One step in the method of reading must be to form concepts on the units of meaning. So it is in science, where a methodological prerequisite to useful observing is to form concepts of things so that tokens of sensation can be grouped into interacting kinds of things and the relevant kinds can be individuated and understood.

Marks on a page, even when individuated and organized, are not helpful for understanding the text unless they are seen to indicate some piece of the plot. They must mean something if they are to be informative. This step from syntax to semantics is a huge one, as is all too clear to someone learning a new language. I can read modern Greek, for example, in the sense of being able to read the alphabet and pronounce the words. I even know enough grammar to be able to identify verbs, nouns, adjectives, and so on, and I am a tireless annoyance to my Greek friends as I love to read out loud all the signs, menus, magazines, or whatever I see. But I haven't a clue what any of it means. Well, not much of it anyway.

The level of information I can get from a Greek text will not be sufficient for translating it into a meaningful message. It is not enough to allow me to understand the plot or to test out any speculations I might have about what is going on in a Greek book. The text is completely opaque without some claims of connection between the marks on the page and the meaning they convey. So it is in science, where observations are meaningful contributors to our understanding of nature only with some claims of connection between the image in sensation and the specimen in the world. To translate the book of nature one needs theories of what the observations mean, a lexicon of what the marks are evidence of.

7

BLURRING THE
INTERNAL–EXTERNAL
DISTINCTION

THE idea of distinguishing between the internal and the external features of theories and regarding these features separately as potential indicators of the truth of a theory was only a start-up suggestion. It has served as a heuristic organizer of the activities of science and the steps of justification. Such organizing concepts are always needed for thinking about evidence, evidence in this case of the methods of scientific justification and of the structure of scientific knowledge. It seemed natural to begin by dividing the fund of information that could be used to evaluate a theory into two categories, depending on the source of the information. There could be information from theories themselves, both the theory being evaluated and other theories relevant to it, and information from the world. Thus justification of a theory would be a combination of theory-to-theory relations, as in the standards of entrenchment and explanatory cooperation, and theory-to-world relations. This matched the internal–external distinction.

The initial suggestion that this is a sharp dichotomy, separating distinctive kinds of features, different sources of information, and characteristic manners of justification, has become untenable in light of the closer look at the relationship between theories and evidence in science. The developing model *of* science is proposed and tested just as a theory *in* science. Just as in science, an initial hypothesis was a helpful background against which to structure the evidence, but the hypothesis is answerable to the evidence and it must be revised or rejected if the evidence demands. In this case, the hypothesis about the structure of science is the claim of dichotomy between internal and external virtues. The evidence it both helps to interpret and uses for testing is the detailed description of scientific theories and their justification. This dialectic between the model of science and the evidence about science, their influence on each other, indicates the need to revise the model at this point.

Consider the evidence. Considerations of external features such as a theory's facility at explaining phenomena and its record of successful prediction, are inconclusive without the complement of internal evaluations. What counts is being the *best* explanation, and this is measured by standards of fruitfulness and theoretical plausibility. Thus internal and external features of theories are inextricably mixed.

But more important than this mixture is the fact of science that the observations themselves bear a theoretical component. Each element in what has been regarded as an external test of theory, each appeal to observational evidence, is a comparison between hypothesis and other conceptually influenced beliefs. Observations, the evidential elements for justifying theories, are themselves assertive beliefs. They are caused by influences from the environment, but they are described and justified under the influence of other beliefs, that is, other assertive claims within our awareness. Observation reports are not so different from theories in that they acquire their conceptual relevance and their justification by their fit into the conceptual, theoretical system. Justifying a theory by comparing it to observation is an internal activity.

The complications that have been described in explanation, confirmation, and observation, the complications that have put justification into a situation of always comparing theory against theory rather than against facts in the world, are not avoidable complications. We point them out not to show what an awful mess science has gotten itself into and certainly not to advocate a way to sidestep the mess and get theory directly in touch with the facts. Rather, these factors of auxiliary hypotheses, underdetermination, and the theoretical component of observation are inevitable. They are in the nature of any attempts to know more about the world than the manifest appearances. The important reason for pointing these things out is to make us aware of the predicament we are in. Only if the predicament is clearly understood can we figure out how to deal with it.

THE PREDICAMENT

The enterprise of justifying scientific theories is in many ways like taking a true–false exam. The result of deliberation over each theoretical claim is a decision whether to assent to it or not or, in a more refined way, a decision as to how likely it is that the claim is true. Each theoretical claim is a declarative assertion about the world, and the business of justification is the issue of truth. Sciences assert, for example, that the ocean's tides are caused by gravity, and the process of justification is engaged to report on the warrant for believing this to be true. All sorts of hypotheses are proposed in science. Cosmic rays are the debris from reactions in distant stars. There are some cosmic rays in the room right now. Modern humans inhabited the North American continent as early as 15,000 years ago. All of these are true or false, and it is the burden of justification to decide which each is likely to be.

In science, no one ever gets a look at the answer key. There is no direct access to information on the truth or falsity of the theoretical claims made. This access problem is clear in cases of claims about unobservables or, respecting the degrees of observability, about things that are only distantly or indirectly observable. Things like cosmic rays, atoms, DNA molecules, or Greek Bronze Age palaces can never be plainly seen. There can be no clear view of such things with the unequivocal news that of this we have spoken the truth. There is at best evidence of these things, evidence, like lines in a cloud chamber or the half-buried ruins of stone and ceramic, that guides the assessment of truth of the theoretical claims. The evidential claims themselves are accounted for by theories, showing that they too are but other entries on the exam and not inside information from the key.

Even evidential claims that are of observable things like the books in the room, the streaks in the cloud chamber, or the basic excavation data of fragmented ceramics and jumbled stones, even these claims are on the exam. These direct observations are not the answer key, the report on simply how things are in the world, for, as we have seen, they are subject to justification and are revisable under pressure from theories. The situation of coming to know about the world doesn't allow a look at the world simply as it is in itself, free of the vagaries of conceptual description. This checking the facts – those things that are undescribed, ineffable but certain, that is, beyond accountability – calls for an impossible combination. That which is ineffable makes no claim and offers no information, certain or otherwise, that can serve as evidence. The theoretical component that all observations must bear to function as useful, informative, accountable evidence makes observations revisable and even rejectable. They are definitely not right off the answer key. Observations too are questions on the exam. Their justification, like that of the more theoretical claims, will be based on their relations with answers to other questions. They will be evaluated in terms of consistency, explanatory relevance, and relations of entailment with other theoretical and observational reports.

The ambition of science, to describe more than is apparent, keeps it steadily at risk of being wrong. There is always an informational gap between the objects of theories and the evidence used to formulate and verify the theories. Claims about the microworld, the very distant world, or the past are motivated and justified by observations of events and objects that are large, nearby, and present. The accessible data are in features of observable phenomena that are caused by the unobservable objects of theory. The image we see of a DNA molecule as projected on the screen of an electron microscope is not the molecule itself, but it is caused by the molecule. Similarly, the tracks photographed in a particle detector or the clicks of a particle counter are not themselves subatomic particles, but they are caused by those particles. The effects bear information of the cause, and the challenge of science is to discover features of the cause through evidence in the effect. Effects are all we have to work with.

Effects are informative of their causes only with the assistance of causal theory. That is, claims about the unobservable causes are products of inference. And the inference must be sanctioned by a theoretical conditional of the form: If the observable effect has feature x, then the unobservable cause has feature y. With evidence and object of theory on opposite ends of this interaction chain, the available data can be useful evidence only in concert with theories to describe the chain and trace the flow of information from one end to another. We are stuck at the effects end of this chain, always interested in the causal end.

However the process of justification in science goes, it must acknowledge this predicament. It must pay attention to what information is available to us and what is in principle beyond the limits of awareness. Only the available information can be relied on to guide the justification and to supply the proof of theories.

JUSTIFICATION AS COHERENCE

Justification is to be a symptom of truth. A properly designed process of justification will allow us to say that if a belief, a theory, is justified then it is likely to be true. The justification will take place either by comparing the theory to the world to see if the description corresponds to the facts or by comparing the theory to other theories to see if it fits into the theoretical system and coheres with the web of theories. The former relation of correspondence with the facts is indeed what we mean by the truth. Correspondence is the nature of truth, but it is exactly this relation of correspondence that we are unable to evaluate. The world side of this theory—world correspondence is inaccessible information. It is the causal end of the interaction chain. It is the answer key we never get to see and so never should rely on for guiding scientific decisions. Much as we would like to use correspondence-with-the-facts as the measure of justification, we can't. We simply have no direct way of telling when correspondence has been achieved.

We are stuck with coherence, judging theories by their relation to other theories. It is not that coherence *is* truth; rather, coherence *indicates* truth. This claim still needs some work, but a system of coherent theories is more likely to correspond to the way things are in the world than is a less coherent system. Coherence, if it is to be the standard of justification of theories, must indicate correspondence.

One thing is for sure at this point: If there is any way to justify scientific theories it will have to be by coherence, by judging the relation between theories. This is assured on the authority of the prerequisite that justification be evaluated on the basis of available information. The method of justifying theories will involve testing by comparison to other theories and by comparison to theory-sponsored evidence. The accomplishment of justification is finding a high degree of coherence among these factors. Thus the process and result of justification are all internal to one's cognitive awareness. Not that it must be grasped all at once by an individual scientist, but the process must be humanly accessible. As science is a communal undertaking, the process of justification, the standards of responsibility in what to believe, must be internal to the community of investigators. Scientific theories, in other words, must be assessed from within the theoretical system of science. There is just nothing for it; we cannot step outside the system for a peek at the facts. And it is not for moral or social reasons that this is the way things are. It's not a matter of pigheadedness or willful bias. It's just the way things are. We must do the best we can with the information that is available, and that means using coherence as the measure of justification and the indicator of truth.

The distinction between internal and external features of theories has been abandoned, and we have settled our attention on the internal side, all for reasons of availability of information. Truly external features are not within our domain of awareness and so are not worth considering when we are looking for useful ways to evaluate theories. The same concern and the same argument force us to abandon the distinction between theories and facts. If facts are the direct information of the way things are in the world, pure and uncolored by influences of description, with no filtering of preconception, then they are beyond our cognitive grasp. If there are such things as facts, they are useless to science because they are unavailable to science. This doesn't mean that there are no such things as facts and that all there are are theories. It means that we cannot know of the facts and so cannot appeal to them in justification. As far as we are aware, there is only theory, and that's as far as we can go in a responsible, serviceable method of justification. The internal–external and theory–fact distinctions are conflated in that anything that is useful to us, anything recognizable and discussible and accountable, is on the internal-theory side. What we take to be facts, observations of the world, are laced with theory.

Denying the fact–theory dichotomy, restricting attention to the theory side of things, is not a suggestion of anarchy. It doesn't lead to a science that is thoroughly subjective, where anything goes and justification is by fiat or caprice. There still are standards, internal standards as discussed in Chapter 2, and there still are internal procedures (more to come on these) to provide guidelines for justification. Though the process must be from within the system of beliefs and theories, it need not be subjective. We can hold out for a standard of objectivity that is complete with the information available, a notion of internal objectivity.

129

OBJECTIVITY FROM WITHIN

Abandon the internal–external dichotomy and conflate the parallel distinction between theory and fact, but don't give up on an objective–subjective distinction. Objectivity can be characterized in two ways, and one of them is amenable to entirely internal evaluation.

One way we could use the concept of objectivity in science would be to describe a statement, like the statement of a theory, as being objective if it is accurate about the object it describes. Objective statements are about real things; subjective statements are about things and properties we make up. This makes objectivity a very desirable feature of science in that it removes the judgment from the control of people. Objectivity, in this sense of the statement being in fact about a real, external object, is good because it blocks any influence of whimsy, prejudice, dementia, bias, or any other human persuasions that steer theory away from truth. It is a wonderfully clear and firm and noble standard. It is also utterly useless. It is warmed-over correspondence, and we have no means of evaluating the link between statement and object in the world.

There is another way to employ the notion of objectivity. The idea is still to avoid the human influences and biases, the subjective factors, that are obstacles to the truth. This aim is furthered by an insistence on objective processes and methods in science. Science is done objectively insofar as it is done with an openness to discussion and a willingness to hear and consider objections to any theory. Objectivity in this sense is in the spirit of putting one's ideas on the line, being open to testing and revision. It is being open-minded and not insular, defending a theory by confronting any challenges rather than by insulating it from challenges. Opening the process of justification in this way removes it from the private and personal influences associated with subjectivity. Requiring an ongoing attitude of openness to challenge isolates the justification process from the personal or social desires of the scientists. It prevents our believing in a theory simply because we *want* to believe it. And the good news is that this objectivity of process, in contrast to the objectivity of a statement, is something that can be evaluated. From within the system we can tell if a theory is testable and challengeable and thereby amenable to an objective process of justification.

This style of objectivity is closely linked to intersubjectivity. Matters of justification cannot be privately controlled. An open, intersubjective process will constrain the individual's influence. It does not, however, constrain the influence and control of the community, and given that scientists band in rather homogeneous groups of similarly trained individuals, it is not so clear that the vagaries of community decisions won't simply reproduce the vagaries of its individuals' decisions.

This concern helps to point out that the concept of internalism can be broken into levels. There are different levels of internalism in the sense that an intellectual process, such as justification of claims about the world, could be restricted to the information available to a particular person, to the information available to the whole community, or even to every person dead or alive. There is a level of internalism that is internal-to-a-particular-theory, meaning that arguments and evaluations of claims are in terms of concepts and standards that are under the influence of that theory in particular. But there is also a level of internal-to-theories-in-general, which requires that our knowledgeable activities be dependent on information within the ensemble of theories we hold about the world. It is this last, most generous version of internalism that is the predicament of science as sketched above.

Thomas Kuhn, in his influential book, *The Structure of Scientific Revolutions* (first edition, 1962), introduces yet another shade of internalism. Because of the high profile of Kuhn's account of science, it is wise to locate his position with respect to the account unfolding here. Kuhn is most concerned with the succession of scientific theories, as from Newtonian mechanics to relativity, or from the caloric theory of heat to the kinetic theory, and the assessment of the new theory as being somehow better than the old. The evaluation of a theory, according to Kuhn, is a process that is internal to what he calls a paradigm. This concept of a paradigm is the key to Kuhn's model of science. The current paradigm of a particular science includes the theoretical commitments of the scientific community, that is, the theoretical claims it endorses and the theoretical terms it uses, as well as the experimental procedures and standards accepted as valuable. For example, physics since the 1930s has been done within what could be called the quantum-mechanics paradigm. Virtually the entire community of physicists conduct their investigation of the world from the perspective of the quantum theory, which prescribes what machines to build, what experiments to do, when they have been properly done, and what the results mean. Physicists speak a quantum language and model the world with quantum concepts. It is their mind-set, their world view, their way of speaking and doing business. It is their paradigm.

Since, according to Kuhn, the paradigm in which scientists work influences the meaning and evaluation of the observations they make, the justification of theories, which includes the comparison of theories with evidence, is internal to a paradigm. Theories can be justified in the context of this paradigm or justified in the context of that paradigm, but there is no chance of justification pure and simple. Thus Kuhn's internalism is somewhere between the narrow internalism-to-a-particular-theory and the broad internalism-to-theories-in-general. It is an internalism to a particular ensemble of theories-plus-experimental-guidelines, internalism to a paradigm.

This level of internalism leads to an interesting and controversial description of how scientists decide to abandon one account of the world and adopt a new one. Whole paradigms, in Kuhn's

132

model, rather than individual theoretical claims, are the units of scientific change. There are no tests of theories that can be done independently of all paradigms, so each paradigm is left to assess its own theories, sponsor its own evidence, and set its own standards. Alternative paradigms, such as the one that embeds the caloric theory and that of the kinetic theory, do not even use the same language and so cannot really be directly compared. They are likely to specify different experiments as the important tests of a theory of heat and to specify different standards and meanings to observation. The scientific allegiance switches from one to the other, according to Kuhn, not because of some objective testing, that is, testing that is uninfluenced by either paradigm, testing that rules in favor of one theory over the other. It is rather a wholesale switch of paradigms including the theories, experimental standards, important experiments, and all, a switch motivated by an accumulation of unsolved problems within the old paradigm and a willingness to adopt the whole of the new paradigm.

Kuhn's point is that the basis for asserting one account of the world over another is not and cannot be objective in the sense of being founded on brute facts that are independent of either account and therefore agreeable to both parties as adjudicator of the issue. There are no such facts, since all information is internal to a paradigm and hence in collusion with one account or the other.

This reasserts the point that objectivity in the form of brute facts in manifest correspondence to the world is beyond our means. It also sharpens the burden on objectivity as an openness to challenge from a variety of perspectives. The burden is to show that the variety can cross the boundaries of paradigms and allow for meaningful tests that genuinely put the theory at risk.

Objectivity of process, in other words, is a nice opening suggestion for a notion of internal objectivity. It is clearly based on available information, but it falls a bit short of our expectation of all that objectivity ought to be, and it is, at this point, a bit thin on detail. We will have further suggestions for the internal concept of objectivity and its evaluation in Chapter 9.

This chapter is short because its message is so simple and yet so important. The kernel of the message is a humble truism: Whatever you use to prove your theories, to justify the belief in their truth, must be information that is accessible in the sense of being within the domain of human awareness. We must be aware, or at least potentially aware, of the justifying factors used to establish the credibility of scientific theories and to keep science on a responsible, truth-conducive method. Justifying factors must be recognizable, discussible, and accountable. If they are observations they must be described. Justifying factors, in other words, will be theoretical factors. We will always be at the mercy of theory. The assessment of scientific claims, the justification of theories, is confined to be done from within a theoretical system. Coherence is the only option, so we had better be sure it has something to do with the truth.

8

COHERENCE AND TRUTH

A LL of this scrutiny of the scientific process got started and has been driven along by the question, Why believe that what science says of the world is true? This is the appropriate time to face up to the question and call on the description that has been built of how science moves to gauge its movement toward the truth. The question is of the accuracy of the scientific description of the world. To put it in a more realistic form and in a form more amenable to constructive answers, ask which aspects of the scientific description are accurate. And, most important as a general evaluative tool, we are interested in knowing a method by which to tell the true from the false. What is it about the process of science that we should attend to as the reliable, responsible process of justification that issues warrant for belief?

This question has been refined somewhat in light of the guidelines that have developed through the description of science in action. From reviewing the practical and conceptual limitations on the activities of explanation, confirmation, and observation, restrictions on the nature of justification have emerged. There is no compromising on the goal, though. Justification is still supposed to be the indicator of truth, or at least likelihood of truth, and there is nothing fancy or philosophical in the notion of truth. It's just plain old truth we are talking about. A theory is true if it is an accurate representation of some part of what there is in the world, that is, of what kinds of things there are and how those things behave on their own and in interactions with us. A theory, like any belief, any proposition, is true if it corresponds to the facts, to the real world. By truth in science we mean exactly the

135

same thing as truth in a court of law. The purpose of a trial and the guidelines of conduct are to find out if the defendant did in fact commit the crime and how. Testimony and verdict are true simply if they accurately report on what happened. Juries and judges don't create the truth by consensus or decree; they must discover it like the rest of us. And so it goes with scientists. The crux, of course, is knowing how to recognize the truth when it comes your way. That's justification.

Though truth *is* correspondence with the facts it cannot be *recognized* by its correspondence. We cannot rely on the facts to guide proofs of scientific theories since the facts are irretrievably at the outer end of the correspondence relation. We could tell that the correspondence link between theory and fact was secure only if we could check both ends, that is, describe and compare both ends. But the fact end, we have learned, is not conceptually accessible in this way, and so relying on the facts is not a useful approach to gathering evidence. The link of correspondence is what we are after but it cannot be directly evaluated, and any useful process of justification must respect this restriction.

So any indicators of the truth of a theory must be internal. That is, they must be within the conceptual and theoretical system of the group of scientists if not of each individual investigator. Justifying factors must be features of which we can be aware. Thus justification must be found in relations among items in the conceptual system, that is, relations among beliefs or, in the more public context of science, relations among theoretical claims. All of the pieces of the puzzle of justification must be recognizable, evaluable, and relevant to the theoretical claims to be justified. The pieces must all have a place in the theoretical system. The process of justifying, then, is a process of comparing aspects of the system, and the accomplishment of justification is the demonstration of coherence among the aspects.

With these guidelines and their genesis in the description of the activities of science, the question of the truth of scientific theories becomes more specific: What does coherence have to do with truth? Why does a coherence among theoretical claims and theory-laden observation claims indicate a correspondence between those claims and the world? Why should links of any kind among theories be a sign of links between the theories and facts? Coherence among theories will secure a cozy network of cooperation and consistent beliefs, but there is no obvious reason to think it will secure any anchor to reality. A requirement of internal coherence, if it is to be a requirement conducive to truth, must be shown to restrict fanciful invention of systems of theories that have no relation to the real world. This is a worry about the possibility of coherent fairy tales, a worry made explicit in 1934 by Moritz Schlick:

> If one is to take coherence seriously as a general criterion of truth, then one must consider arbitrary fairy stories to be as true as a historical report, or as statements in a textbook of chemistry, provided the story is constructed in such a way that no contradiction ever arises. [Quoted by Israel Scheffler in *Science and Subjectivity*, p. 102]

It is a legitimate concern that coherence is too easy, too liberal a standard to be the mark of that unique system of theories that is the true account of the world. It is a worry to take seriously.

THE DIALECTIC BETWEEN THEORY AND OBSERVATION

It is important to consider the activity of justifying scientific theories, the endeavor of increasing the overall coherence of the theoretic system, as an ongoing process. The scientific description of the world is not a static system of claims, and the point of justification is not to show that the current snapshot showing all of today's theories shows nothing but the truth. A model of justification must acknowledge the changes occurring in a theoretical system. Theories come and go and are altered, and most

137

importantly, observational beliefs are steadily added and must be acknowledged and dealt with. The changes, if things are working properly, will be generally in the direction of overall coherence. That is, decisions on what changes to make in the system of theories and on how to account for and accommodate new (and old) observations are made with a mandate to free the system of inconsistencies, to link components and sections of the system by explaining some claims in terms of others, and generally to show that seemingly disparate parts of the picture fit together. The achievement of microbiology, as an example of a recent and significant change in the theoretical system, was to demonstrate a harmony and relevance between the lifeless claims of chemistry and molecular physics and the network of theories and observational claims describing things biological, genetic, and evolutionary. By bridging these domains and establishing the relevance of chemistry to life, microbiology scores high on the virtue of fruitfulness. It promotes coherence in the system by stitching together a great variety of phenomena with a minimum of theoretical thread.

The changes to a theoretical system are not only to promote a neater fit among theories but also to accommodate and influence the accumulating observations. This is the dynamic stimulus to the system and the greater challenge to the maintenance of coherence. The task of science is to deal with the input of observation and revise the system of theories as necessary in the direction of greater coherence. The question to be asking at this stage of the analysis is whether the direction of coherence is the direction of truth. In the ongoing process of justification by coherence, what reason is there to think that science is headed truthward?

The justification process in science clearly pivots on the activity of observation. Even with the burden of theory and conceptual influence, observations are still the link to the environment and will be, if anything is, relief to the worry about coherent fairy tales. Observations must play an active and influential role in blocking any significant warrant for belief in creative, artifactual systems of theories that purport to describe the world but in fact are wildly inaccurate.

This is speaking as if there is a significant difference between observations and theories or between observational claims and theoretical claims, but the work done in Chapter 6 was to show the similarity between theory and observation. This calls for some clarification. Theories and observations are significantly similar in terms of their accountability. Both kinds of claims or, if unspoken, both kinds of beliefs, theoretical and observational, are *of* something, and this of-ness, the descriptive, conceptual meaning of the belief or claim, is based on a relation to other claims, beliefs, and theories. The descriptiveness of a theory or an observation is secured by its fit into the current conceptual network. In this important sense, observations and theories are the same sort of thing.

They are alike as well with respect to their justification. Neither theory nor observation is admissible into the system of science without proper credentials to indicate its accuracy and authority to report what it claims. And, for both theory and observation, that authority is based in a cooperation with other components of the theoretical system. Theories are explained, implied, or sanctioned by other claims in the system. Observations are supported by accounting theories, that is, by other claims in the system. In this other important sense there is no distinction between observation and theory.

In another sense, though, there is an important difference between theories and observation. That is in the genesis of the beliefs, the source of the ideas. Theories and theoretical claims are the result of reflection. Their origin is in thought in that they are produced through inference or suggestion from other ideas, other claims. The idea that the sun maintains its supply of energy from an ongoing process of nuclear fusion at the core is an idea we come by through thinking about things, putting two and two

139

together, so to speak, to see what our other beliefs about energy and astronomy suggest about solar energy. The claim about solar fusion is not caused by any *particular* sensation, though in their lineage many of the claims from which the theory about the sun's energy has been inferred will themselves be linked to sensations. But the claim about the sun's source of energy is one that can be thought of with our eyes closed. It is the pure product of the activity of theorizing, of thinking about, following suggestions and implications of other ideas.

Contrast this with beliefs that enter the conceptual system in a spontaneous way. Many beliefs come to mind not from an active process of inference but as the direct causal result of interaction with the world. Observational beliefs are each one caused by a particular sensation. Their assertive content, that is, the conceptual description that provides a handle on what the sensation is of, and their accountability as reliably caused beliefs are established in relation to other beliefs and theories. The spontaneous beliefs from observation must be dressed in theory to participate as evidence, but they need not be motivated by theory. Observations are forced upon us by the environment. They are the spontaneous effects of particular sensations.

Recall the cosmic rays that present themselves as streaks of vapor in a cloud chamber. A glance at the chamber, crisscrossed with fleeting vapor trails, may give us the idea that there are cosmic rays raining through the room. Describing the observation as being of cosmic rays, associating that descriptive concept with the sensation and sanctioning the view as an accurate source of cosmic-ray information, is a matter of fitting into a web of theory. An appeal to accounting theories describing the causal, informational link between a cosmic ray and the visible track and describing the conditions for viewing the track is what licenses the report as useful evidence. This accountability indicates that what we come to believe by looking at the cloud chamber we believe truly. But we come to believe by looking. We got the idea about cosmic rays in the room in the first place as a result of the particular sensation. The belief is caused by an influence from the world, not by an active process of thought. It was a spontaneous belief, an observation.

This marks an operative difference between observational claims and theoretical claims, between observation and theory. It is a difference of degree since many scientific claims are the product of reflection about particular sensation and so are hybrids between observation and theory. There is nonetheless this useful distinction based on the cause of a belief rather than its descriptive content or source of accountability. Observations are the result of causal, sensory input from the environment. The objects and events in the world do not directly justify the scientific claims we make, but they do cause many beliefs, those that are the spontaneous results of particular sensations. This is the distinction and the importance of observation.

The evidential authority of an observation is based, as has been said, in its accounting theories. Given the pivotal importance of observation as evidence, it is clear that the justification process wants the best accounting theories it can get. Much of the weight of the question of believability of science rests on the accounting theories. Why believe *them?* In fact, the work of analyzing justification has only been postponed to this point and focused on this link between observational evidence and the theories for which they stand. How are the accounting theories confirmed? Like any other theory.

Accounting theories are nothing special. They are distinguished by their manner of use, not by their content or degree of generality and not by their manner of confirmation. Any theory can be used in the role of accounting. In one context a theory may be the valued end, the accomplishment of description. Theoretical claims in optics, for example, are simply nice to know. Light and its behavior are part of the world, and so optics is part of the accomplishment of science. But the theories of optics are also nice to use. They function well as accounting theories that keep track of the informational flow from the environment to ourselves. The same is true of theories of neutrino physics. They are useful as accounting theories, describing observations of what is happening in the solar interior, and they are in their own right informative theoretical descriptions of an aspect of the world. Their being *accounting* theories is a circumstance of their use on a particular occasion. So asking how an accounting theory is confirmed, asking how we know what we do about neutrinos, for example, is exactly a return to the general question of confirmation.

Accounting theories, like all theories, are confirmed by their plausibility relative to other theoretical claims and by their coherence with observational claims. Theories of optics or neutrino physics, active players as accounting theories, are confirmed in the familiar pattern of explaining phenomena and making successful predictions. Theorizing about light waves and their interactions explains all sorts of common phenomena from the colors of the rainbow to your short, fat legs as you stand waist-deep in a swimming pool. Theorizing about neutrinos explains what happens to the energy in some otherwise mysterious interactions between elementary particles. These theories, employed here and there in support of observations, have their own explanatory and predictive successes. They too confront the observational data and thereby establish their justification.

The general picture that has emerged of justification in science is one of mutual support between observations and theories. Theories, whether or not they find a use as accountants, are understood and justified by reference to observations. But then, observations are understood and justified by reference to theories, accounting theories in particular. That is, observations are accountable to theories. Neither sort of claim is the single, basic source of justification in science. It is rather that the justification derives from a mutual support between the two. This may sound

vaguely circular, and it is vaguely circular. But it is not a problematic circle in the sense of immediately impeaching the process
or the evidence. To see why the circle is tolerable (as well as inevitable) it is better to think of the relation between theory and
observation as one of negotiation.

Theory and observation, in the activities of explanation and
confirmation, engage in mutual justification. It is a process of
give-and-take between the two, of adjustments to either side, but
always under the constraint to increase coherence. Scientific
theories report the big picture of the world. They provide a global-
scale understanding of what sorts of things there are in the world
and how they interact. This understanding then allows us to
make sense of sensations. Observations, given meaning and credibility by theory, report on individual situations in the world and
provide a more local-scale description. The observations then
constrain theorizing since theories must be compatible with observations. And more, theories must explain and show cause
for observations. Even with the complication and ambiguities
inherent in explaining and confirming, with the necessity of
auxiliary theories and questionable conditions of testing, in all
cases of comparing theory to observation the result must be compatibility. Among all the participating claims, the theory being
tested, the auxiliary theories, the observation claims, justification
requires a thorough compatibility. If a prediction or explanation
fails, that is, if theoretical claims are in conflict with observation,
something has to give. The goal is coherence, and it begins by
purging inconsistency.

The influence between theory and observation goes back and
forth. It is not always in the same direction, say, observation forcing the ways of theory. With a steady input of new observations
the system of scientific claims must be kept open for revision to
insure a fit for new observation claims. These new observational
beliefs cannot be ignored. If they cause trouble by contradicting
or offending or casting doubt on theories in the network, then a

readjustment of the network or of the observation itself is in order. You do what it takes to move the developing system of claims in the direction of greater coherence. You do what it takes to tighten the explanatory links between claims. And you do anything it takes to free the system of inconsistency.

This is the most general view of the process of scientific justification. It is an ongoing process of accommodating new observations within the system of theories and doing so under the constraint of enhancing coherence. It is surely not a monotonic rise in coherence that takes place in real science. We may pay with incoherence in one part of the system for a tighter fit somewhere else. Or we may provisionally accept new ideas that bring disarray, on the hunch that they will eventually precipitate a greater coalescence in the system. No one expects each step to be an enhancement of coherence, but there is good reason to expect a long march to be generally in this direction.

A consequence of this processual view of science is the expectation that individual theories and whole patches of related theories can change over time to allow an increased coherence in the larger system. These changes could result from the acceptance of new observations that threaten old beliefs. Changes in the theoretical system could also be the result of clever thinking about old ideas, discovering ways to find agreement between disparate patches within the system. That is, it is not only observations that can cause a reshuffling of theory; the recognition of imperfections within the system of theories can do it as well. Our current network of theoretical claims about the world is not perfectly coherent. It is modular in the sense of having several seemingly unrelated patches, and it is likely always to be this way. There is a nest of claims about relativity and gravity, for example, that has nothing to do with the group of claims about the evolution of life. There are even collections of theoretical claims within the system that appear to be inconsistent. Quantum mechanics and the theory of relativity, two of the best and brightest in physics

144

today, do not always get along. As described by general relativity, the material at a spacetime singularity (as described in Chapter 1) must be smashed to a single point, infinitely concentrated and precisely located. A quantum-mechanical description of the physical world explicitly prohibits any such point locality of material stuff. So what are you going to do? Even with no new observations to consider it is clear that the theoretical system wants adjusting for consistency and explanatory closeness, which is the achievement of coherence.

Einstein's development of special relativity is a nice case of changes to the theoretical system with the motive and result of reconciling differences between theories and enhancing the general coherence. The central idea of relativity, that no frame of reference is distinguished as being stationary in the universe, that all frames, in other words, are created equal, was certainly not new with Einstein. What was new was the adjustment to the theoretical system to smooth the inconsistency between this principle of relativity and the new but phenomenally successful theory of electromagnetism. The principle of relativity insists that no possible measurement can be performed that would distinguish one uniformly moving reference system from another. The equations of physics are exactly the same on a moving train as on the stationary ground. The theoretical account of electricity and magnetism, though, seemed to suggest that the effect produced by moving past a stationary magnet would be different from that produced by moving the magnet past a stationary observer. The electromagnetic theory, in other words, distinguished a preferred frame of reference, namely, the one in which the magnet is stationary.

The accomplishment of special relativity is to reconcile this difference by changing the notions of time and space, the media of measurement, to fit the theory of electromagnetism into the framework of relativity. By saying that the measurements of space, time, and mass are dependent on the relative motion of the measurer, Einstein sacrificed entrenchment to resolve an inherent inconsistency and to enhance the coherence of the theoretical system.

Einstein's revisions to the understanding of measurements of space and time also allowed the theoretical system to accommodate the new observations of the Michelson–Morley experiments. Michelson and Morley, at about the same time that Einstein was thinking up the special theory of relativity, devised and performed an experiment to detect the effects of the earth in its travels around the sun, moving through the special rest frame as suggested by the electromagnetic theory. They found no effect. The experiment was repeated under various conditions, and in various locations to the extent that its null result could not be blamed on experimental conditions. The observation, or the expected observation, was certainly heavily theory-laden. It's not that a sufficiently sensitive observer feels a breeze as the earth plows through the sea of ether that rests in the electromagnetic frame, nor can we see any movement relative to stationary points of reference. To detect the motion one splits a beam of light into two and sends the two beams at right angles and notes if one has taken a slightly longer time to travel an equal distance, longer because it was slowed by fighting the current of the ether. The comparison to travel times is done by reflecting and recombining the two beams and noting the bands of light and darkness that are caused by the wave properties of the light and are influenced by the relative arrival times of the two waves. It is not a simple experiment, and the observation that there is no motion of the earth relative to the ether depends heavily on accounting theories to link the final fringes of light (the things we observe) to information about the motion of the earth (the event we are interested in). In fact, attempts were made to accommodate the refuting evidence by blaming the accounting theory. The effect (the earth's motion through the ether) is really there but we are just not seeing it with this Michelson–Morley apparatus because

146

the pressure of the ether on the leading edge of the table holding the optical equipment actually compresses the table so that the light that must fight the current has a shorter path to travel. The shorter path at slower speed just compensates, and that is why the two beams of light arrive at the point of detection at the same time.

This bit of "ad hocery" never really caught on. But in any case, it is clear that the Michelson–Morley observations could not be ignored. They forced a negotiated agreement among the observation report, the accounting theories, and the theory being tested. Einstein's theory of relativity suggested a change in that part of the electromagnetic theory that mentioned or required a preferred reference frame or a medium such as ether to support the electromagnetic effects. It modified as well scientific beliefs about measurement in space and time. The net effect was that a null result of the Michelson–Morley experiment was to be expected. It's not that Einstein's theorizing was motivated by or in response to the Michelson–Morley experiment. Rather, both the theory and the observation gain credibility by their compatibility and by the newly demonstrated harmony between the theoretical description of electromagnetism and the principle of relativity.

The point is that an applicable model of the process of justification in science must have room for this aspect of change in the theoretical system. The change, as in the case of special relativity, is swept by the tide of coherence. In the case of special relativity, basic consistency among theories is gained by the introduction of novel and rather startling (a polite way of saying prima facie implausible) ideas of the nature of distance and time. As the network of theories changes and as new observations are accommodated, justification changes. The special theory of relativity replaces the Newtonian description of motion, but that does not mean there was never justification for Newtonian mechanics. Nor does it mean we will always be justified in the special relativistic mechanics. Under the circumstances that preceded the

Michelson–Morley experiments and the clear development of the description of electromagnetic radiation, the Newtonian view was justified. But under current circumstances, that is, in light of the current network of theories, the relativistic view is justified. Scientific justification must be contextual justification, sensitive to the circumstances of the time. The alternative is to put off all evaluation of justification until the end of science, whatever that means, and see what stage in the development showed the greatest coherence. But that is doing nothing at all.

The process of science presents a steadily developing picture of the world, a picture whose coherence is maintained and even enhanced by an ongoing negotiation between theory and observation. Stages of the process are often begun by speculation, taking a guess about what the world is like, to break into the cycle of theory in conceptual support of observation and observation in evidential support of theories. Hypotheses, the currency in the context of discovery, are needed to get started since a theoretical framework of some kind is necessary to do anything worthwhile, to think, to observe, to test. But the hypotheses used to break into the cycle are not irresponsible, carefree speculation. They are accountable. They must be tested for coherence with other theories and with observations. They and other theories are subject to revision and are used to account for observational evidence, which in turn directs revisions in the system and perhaps the suggestion of new theories that will influence and be influenced by observations.

READING THE BOOK

This process of negotiation between theory and observation and the account of the developing theoretical picture of the world bring out in full the methodological similarity between science and reading a text in an unfamiliar language. An unfamiliar text

148

can be deciphered by a process often referred to as the hermeneutic circle. It is **hermeneutic** in that its goal is interpretation
of the text, and it is circular in open acknowledgment of the necessity of using only our available information to account for and
extend our available information. It is circular because of its limited resources. There is only the text, the marks on the pages. We
have no access to the actual events being described, assuming it
is a historical narrative of events long past. Nor can we interview
the author and get the straightforward explanation of the plot.
Whatever we make of the text must come from what is on the
pages and from our own theorizing. Initially, of course, the marks
on the page are meaningless to us, and the larger message of the
text is unknown.

The hermeneutic method of interpretation is very similar to the
scientific method of understanding the world. By describing this
similarity I do not mean to suggest that it is anything more than
a methodological similarity. That is, there is no suggestion that
the objects of natural science are meaningful or symbolic as are
the linguistic marks of textual exegesis. Nor is there any hint that
nature must have an author as does a text. Understanding a book
involves understanding the intentions of the author, and intentions are particularly illusive. The meanings of words, as we
know from Chapter 1 and from our own experiences, are under
the control of people, including authors and those who influence
them. There is no tight connection between the words printed on
the page and what the author intends them to mean. It is a connection that is influenced by the cultural and intellectual environment of the author, and a connection that is not immune to
the vagaries of the writer's caprice and whimsy. These complications must be dealt with in translating texts, as they are under the
control of people with who knows what on their minds. These
complications are missing in natural science. Nature has no author and does not have to deal with intentions behind the events

it understands. The objects of study by natural science and by textual interpretation are very different, but the method of study, as we will see, is significantly similar.

Consider first the hermeneutic method of interpreting a text; then we will see how the method is similar to the process of science. To translate the text one must initially speculate on the meaning of individual words and passages, speculation based on patterns of recurrence of symbols and on their place in the context with other patterns and passages. These initial speculations on the meanings of the words provide preliminary hypotheses about the messages of individual passages. They also lead to a way to test the speculation. We can try out the assumed assignment of meanings on other occurrences of the symbols, other passages in the text, and see if the resulting translation makes sense. If it does, the standards of testing can be tightened by asking whether the newly translated passage is consistent with others that have been tentatively translated already. Stricter still, do the passages as given meaning under the speculated translation hang together? Do they show a continuity and relevance to each other that one expects of a coherent text? If not, if translations of new passages are just gibberish, or they contradict other passages, or the whole thing is turning into just a bagful of unrelated, irrelevant claims, then revision is called for in the initial guess at the assignment of meanings to words, or in the speculative theory about the message of the plot.

The process of translation advances by a back-and-forth exchange of information between the developing understanding of the plot and the translation of individual passages. The global understanding, the message of the whole work, guides the local understanding of the parts, the individual passages. These passages are translated, they acquire their meaning, by their fitting into the bigger picture and making sense. The meaning of the parts is discovered from their context. The understanding of the parts in turn guides the understanding of the whole in that the message of the plot is built from the component messages of the individual passages.

This mutual guidance between the local reading of passages and the more global understanding of the text results in an ongoing negotiation between the two. The reading of the text is constrained by a requirement of coherence. It must, in the end, make sense. The passages must be consistent and should hold together in a cogent message, at least in sizable sections of the text. As the reading continues, new passages are encountered and must be accommodated within the network of beliefs about the book and its message. Each new passage is like a new observation, of which the reader must make sense and which must be fit coherently within the theoretical system.

It is the coherence of the interpretation that makes it believable. If the developing understanding of the message of the book is kept consistent and coherent, there is reason to think that it is developing toward the correct reading. Consistency is certainly a necessary feature of an acceptable, that is, likely-to-be-true interpretation of the text. The standard of consistency then can be used to rule out unacceptable readings. It is a way to falsify some suggestions of the meaning of the text as not-possibly-accurate. But consistency alone is far from sufficient. There will be plenty of consistent readings of the book, alternative assignments of meanings of the symbols that produce alternative messages from the text. Sufficient cleverness and diligence on the part of the reader could produce any number of consistent readings of the text, but only one of them will be the correct reading that conveys the intended message (still talking about a book, not nature). To trim this list of potential readings we must apply additional criteria of acceptability, criteria that can be judged on the basis of information found within the text itself and within our own growing understanding of it. The next steps to justify a proposed reading cannot involve consultation with the author or a glance at the events the book is about. These facts are inaccessible, and we must deal with the information as filtered through the text and as organized by our understanding of the text. That is all we have to go on.

Why is consistency a requirement of acceptability of a theory about what the text says? Why aren't inconsistent readings considered? The answer is based on an assumption that the text makes sense. It is a principle of charity, giving the author the benefit of the doubt on the issue of reasonableness. The book, it is assumed, has been written to be informative. It simply does not deny on one line what it asserts on another. The message is free of contradiction, and so our reading of it must be as well. This is an assumption about the way any book *must* be written. It is not something that can be proven through reading, that is, through observations of texts. It is part of the conceptual preparation we must do to facilitate reading. But like every other claim we make about the world or about the text, this one is bendable. As a last resort, if no consistent reading is developing after diligent hermeneutic work on a text, it may be that contradictions have been planted in the message itself. The inconsistency might be just the way things are in the plot. This option has perhaps occurred to many people trying to read *Finnegans Wake*, as good an example as any of a book written in an unfamiliar language. This has to be, though, the very last resort.

Other standards of justification of proposed readings, that is, other criteria by which to cut the field of consistent readings, are generated by extending the principle of charity. The plot of the book was written not only to be free of contradiction but also to hold together in such a way that many of the passages will be relevant to each other. The plot, we assume, makes some sense in that one thing leads to another and what happens here may explain what happens there. The message, it is assumed, has this feature of organization, at least in sizable sections or chapters, and any accurate reading of the text must reflect this through a coherence, a fitting together of the interpreted passages. This criterion of coherence in the reading of the text is thoroughly evaluable with the information on the page and in our theories of meaning. It is an internal feature of these theories, and so it is a useful criterion to further filter the likely-to-be-accurate readings from the less likely.

In the context of science, where the goal is not to read a text or to comprehend the intentions of an author but to describe nature, a similar principle of charity is appropriate. The dialectic of observation and theory proceeds under an assumption that nature cannot tolerate contradiction. Objects in the world cannot both have and not have particular properties. They cannot both be and not be particular ways. Things are one way or another. This is a requirement of how any world must be, and so overall consistency is a necessary condition for acceptability of a theoretical system.

Nature is more than just free of contradiction; it is organized. Events in the world do not just happen, one damn thing after another. Objects interact in consistent ways, and events at one time and place will be relevant to events at another. What happens here and now may influence what happens there and then. This organization in the world must be reflected in our theories of the world and it is, as coherence. Coherence in the theoretical system serves to filter the likely-to-be-accurate theories from the less likely. In this sense the feature of coherence is truth-conducive. As in the case of reading the text, it may not be warranted to assume that the world is wholly organized. There may well be a disunity of things such that organization and relevance are to be found only in pockets, in domains of objects and events. Our theoretical system then may show a modularity of coherent networks but no global cohesion.

Consistency and coherence in the theoretical system are necessary features of an accurate description of the world. But still these criteria are not strict enough to identify a single theoretical description as the true one. Back in the context of reading a text, there will always be multiple readings and multiple assignments of meanings to symbols that are equally consistent and at the moment equally coherent. Coherence and consistency together are not sufficient to rule out all but the correct reading or, in the case of science, the true account of the world. But this is old news. From the beginning we have been warned that there is no hope of certainty, whether it is in evaluating the correct reading of a text or in justifying scientific theories.

The responsible thing to do in this situation is, first of all, to continue reading. In the context of science this means to continue observing. As more passages must be fit into the plot or more observations must be accommodated, the theoretical system will be forced to adapt to an increasingly higher degree of difficulty of achieving coherence. This will serve to further weed out misfit claims and previously, though no longer, coherent theoretical systems.

The other responsible thing to do is to devise other internal standards by which to judge theories or readings. For example, a false reading of the book or a false system of theories of nature would most easily maintain its consistency and coherence if each particular theory was tested by observations that were interpreted by that theory itself. The tight, insular circularity of acting as the accounting theory of your own evidence would severely limit a theory's exposure to possibly refuting information. Realizing this we can count it as virtuous of a theoretical network if the observations used to test a particular theory depend only on accounting theories that are independent of that theory being tested. Even though all evidence must be accountable by theory, that is, it is all theory-laden, we can insist on *independently* accountable evidence. The observations must be laden with theories that are independent of the theory for which they serve as evidence. This restriction blocks the collusion and close,

154

self-serving circularity of a theory prescribing its own evidence. The standard of independence is a significant addition to the list of internal virtues of theories. It is significant enough, in fact, to warrant its own section.

INDEPENDENCE

Independence is another virtue along with explanatory coherence, entrenchment, and the others. Paying attention to the independence of the evidence used in science means we must always get beyond the slogan that observation is theory-laden and in each case attend to *which* theories are used to influence the observations. It is important to the evaluation of justification to identify the accounting theories in each case and see if *they* have a stake in the outcome of the observation. Here is an example: Theories of optics are used to account for the image formed by an optical microscope. Knowing how light behaves in its interactions with the specimen and the lenses tells us how the image is formed and that the image reliably displays features of the specimen. If the image is of a chromosome, that image may function as evidence of some claim about relevance of shape to function of the object or about relative numbers and sizes of chromosomes in different organisms. The claims about optics that are used to describe the formation of the image and to thereby sanction it as an accurate representation of the specimen are independent of the claims about biology and genetics for which the particular observation functions as evidence. The optical theories do not gain or lose credibility on the basis of the outcome of biological experiments. The accounting theories in this case are clearly independent of the theories being tested by the observations.

Why is independence good? That is, why is this feature of independence between accounting theories and the theory being tested a truth-conducive feature? The goal of validation of scientific knowledge is the elimination of theoretical artifacts. We want

155

the entities and events described by a natural science to be features of nature and not simply creations of theory without correlates in the world. The chances that several unaffiliated theories will cooperate to manufacture artifactual evidence are less than that a single theory will do just that.

An analogy can help motivate this point. Newspapers make descriptive claims about the world, as does science. If I read in the paper a story about events that are inaccessible to me, events in China, say, what makes the story believable? In part, a story is credible if it appears in other media such as independently owned and managed newspapers, radio, and television. We are dubious of a story that is reported by only one newspaper and, to borrow from Wittgenstein, there is little to be gained in terms of objective confirmation by buying and reading another copy of the same newspaper. The point is that the chances of a single newspaper reporting falsehoods, by design or by mistake, are greater than the chances of several independent media reporting the same untruths. The quality of the evidence for a story (or a scientific theory) is measured in part by its independence.

The benefits of independence can be further appreciated by considering our own human perceptual systems. We consider our different senses to be independent to some degree when we use one of them to check another. If I am uncertain whether what I see is a hallucination or real fire, it is a less convincing test simply to look again than it is to hold out my hand and feel the heat. The independent account is the more reliable because it is less likely that a systematic error will infect both systems than that one system will be flawed.

Another analogy to the concept of independence is found in history. Ancient history in particular is fortunate to have available two independent sources of information, the literature of ancient texts and archaeological artifacts. Interpretation of artifacts, their age, their use in trade, and the like, is credible insofar as it is consistent with interpretations of ancient texts. In this way, claims

156

made by archaeologists can receive an independent corroboration by the claims made by interpreters of ancient texts.

There are plenty of good examples in the physical sciences of independence functioning as a rational measure of belief in scientific claims. One of the best is in the experiments of Jean Perrin in the beginning of this century, experiments previously mentioned in Chapter 5, designed to measure the number of molecules in one mole of material, Avogadro's number. (A thorough account of Perrin's accomplishment is to be found in Nye, *Molecular Reality.*) Perrin measured the same physical quantity in a variety of different ways, thereby invoking a variety of different auxiliary theories. And the reason that Perrin's results are so believable and that they provide good reason to believe in the actual existence of molecules is that he used a variety of independent theories and techniques and got them to agree on the answer. The chances of these independent theories all independently manufacturing the same fictitious result are small enough to be rationally discounted. It is the independence that supports the credibility of the account.

The benefits to justification in these cases come from the features of independence and coherence. They are not inherited from any particular theory that contributes to the process. Using theories of optics to account for use of a microscope to observe chromosomes is an example of good, credible evidence, not because the theories of optics are well confirmed but because they are independent from the biological claims being tested. The same is true in the case of using carbon-14 or dendrochronology techniques to determine the age of archaeological artifacts. These techniques supply credible evidence, but not because good, hard sciences are allowing their authority to trickle down to more supple science. More importantly, the evidence is credible because its gathering is independent of that for which it serves as evidence.

With the requirement of independence we can ease what may have appeared to be a tension between two key components in the description of science, the facts that observations are influenced by theory and that these observations are the evidence used to test theories. This relationship need not be circular as the theory that does the influencing need not be the same as the one that takes the test. This circle-blocking independence is a measure of objectivity of the evidence and of the process of justification. Independent evidence is outside of the influence of the *particular* theory it serves, though it is still within the theoretical system of science. It is internal, as it must be, in the latter sense, though external of the particular claims it tests. Furthermore, the evaluation of independence is not influenced by psychological, social, or aesthetic factors. It is objectively measured, out of the control of whimsy and bias.

Independent evidence is objective evidence, and the requirement of independence is a key ingredient of the scientific process that prevents problematic circularity in the justification of theories.

9

OBJECTIVE EVIDENCE

T HE image of objectivity has a lot to do with our generally
high regard for science and things done scientifically. Objectively tested claims have a reasonable chance of being true because it is not our personal judgment that makes the call of
justification, so it must be nature that does the adjudication. There
are two possible sources of information for our ideas, ourselves
and the world. Standards of objectivity are intended to quiet,
though not silence, the first in the process of justification, in
hopes that the second will be heard and understood.

Granting this important role to objectivity is fine as long as we
understand what objectivity is or, perhaps more important, what
it is not. Using the concept as a stick to beat down theoretical artifacts or as a banner to advertise warrant to believe that a theory
is true demands an awareness of the limitations of objectivity as a
useful, recognizable concept. It does not help the cause of validation in science to appeal to an idea of objectivity as reference to
external objects or events. It does not help, for example, to claim
that the concept of "electron" is objective just in case there *are*
such things as electrons, because the truth of the claim that there
are electrons is exactly the sort of decision that an appeal to objectivity is supposed to help. Nor does it get objectivity's work
done to describe it simply as intersubjectivity. The concept of
"electron" is objective by this standard or, better, claims about
electrons are objectively tested to the extent that there is general
agreement about those claims among the relevantly trained and
interested scientists. This locates the credibility of science in its
open and public manner of scrutiny. Recent claims about cold

nuclear fusion were put to the test by the community of experts and found to be insufficiently warranted. This is science at its objectivity-as-intersubjectivity best, and this standard is effective at filtering out personal oversight or bias. It blocks from science seeing simply what one wants to see. But it does not block the influences of communal oversight or bias. It does not block the effects of homogeneous training of the community or of the institutional control of publication of new ideas. This harmonious choir of the human influence on justification is not quieted by a standard of intersubjectivity. Only the soloists are discouraged.

The standard of intersubjectivity does, however, have a lot going for it in that it captures the spirit of accountability and openness to review and revision that we expect of objectively held beliefs. In the same spirit of openness to general review and to challenge is the notion of objectivity as intertheoretic review. This characterization of objectivity is the one we will focus on and apply to the process of scientific justification. Unlike objectivity as reference to external object, this feature is recognizable and determinable. And, unlike objectivity as intersubjectivity, this feature devalues the opinions of people and emphasizes instead standards of evaluation as forced on us by our diverse theories and the requirements of coherence and consistency.

THE INDEPENDENCE MOVEMENT

The focus of this useful standard of objectivity is on the evidence used for testing theories. Objective evidence is independent evidence, and independence is a key ingredient for what makes science scientific. It is not the only symptom of good science. Surely science is enhanced by great creativity in the process of discovery and by such features as generality of theories. But when it comes to justification, the crucial feature of responsibility in choosing theories is the insistence on independent evidence.

This suggests a change of focus from observation as the foundation of scientific credibility to *independent evidence* as the key to justification. It's not that observation is not important; it is rather that we are seriously misled by thinking that science is scientific simply because it is based on the facts of observation or that a theory is believable simply because it is tested by observation. Both the process and logical form of testing and the nature of observation are more complicated than these claims suggest. Observations, don't forget, speak as theory and are not so simply the foundations upon which theory is built.

Better to locate the credibility of science in the requirement of *independent* evidence rather than observational evidence. This causes much less worry over the difficulties in sharply drawing the dichotomy between theory and observation or between what is observable and what is unobservable. It easily concedes the fact that the mark of theory is everywhere in science, on all the descriptive claims and on all the evidence. But it shows how to deal responsibly with this predicament. All evidence in science bears the mark of theory, but it does not have to be the mark of the particular theory being tested. Insuring that it is not is the accomplishment of objectivity in justification.

This also suggests a change in the framing of the question of scientific realism. In Chapter 5 we were moved by the tide of tradition to put it this way: Is there a distinction in the justification for scientific claims that matches the distinction between observable and unobservable entities? In other words, is a claim justifiable only to the extent that the things it describes can be observed? This way of putting the question has made it difficult to answer, largely because it has been impossible to determine the base distinction between what is observable and what is not. We certainly cannot decide whether justification is linked to *this* dichotomy if the dichotomy itself will not come into focus. Better to put the issue of realism in this way: Is there a distinction in the justification for scientific claims that matches the distinction between independent and nonindependent evidential support for the claims? A clear question allows a clear answer, and the answer to this one is yes. You can call this a position of antirealism with respect to nonindependently evidenced theories (since we are not justified in believing what they say), or you can call it realism with respect to independently evidenced theories (because we do have justification for these). It really doesn't matter. What does matter is that we pay attention to independence as a significant factor in justification.

It is likely that this feature of independence between the theories sponsoring evidence and the theory benefiting from the evidence comes in degrees. It is not an all-or-nothing evaluation. Neither is justification. Some claims are more justified than others. To see how these degrees of independence manifest themselves we need to look at some examples of the gathering and use of evidence and to attend particularly to the measure of independence.

But first, note that the appeal to independence as the mark of objectivity exploits the disunity of science. At any time, including the present, the system of scientific theories does not show a global coherence. It shows instead pockets of coherence, clusters of theories that are mutually linked through inference and explanatory relevance but that can be revised and even rejected without significantly influencing other clusters. These subsystems of claims make contact, if at all, through the evidence. They intersect on observational claims in that one may account for the evidence of another. It is not just whole sciences like biology, physics, geology, or archaeology that make us these pockets of coherence. Subdisciplines can cluster as well. Within physics, for example, theories of optics are conceptually isolated from theories of gravity. This is why a look at the stars, even when mediated by an optical telescope and imaged on a photograph, is good, objective evidence for the general theory of relativity and its account of gravity.

Note also that this appeal to independence as a measure of good evidence matches our intuitive, commonsense notions of what is good science and responsible reasoning. Imposing the standard of independence blocks circularity in the use of evidence to prove theory and thereby prevents a theory's self-help of providing its own proof. It is the same standard of justification that requires that a court of law listen to more testimony than just that of the defendant. The defendant cannot be trusted to proclaim his own guilt or innocence. Nor can he be trusted to authenticate the material evidence. "That's not *my* gun!" To get at the truth we must hear from independent witnesses who have no vested interest in the verdict. The jury cannot see the crime itself, but independent testimony, objective testimony, allows them to reconstruct the scene on the basis of what they can see and hear. Independence in the testimony, along with consistency and coherence, are the guidelines of credibility.

These intuitions about trial by jury are a reliable guide to scientific justification. A few examples now will show that what we naturally take to be good evidence and unproblematic confirmation is in fact also independent evidence. In the easy cases, in other words, the suggested analysis agrees with common sense. This should enhance our confidence in the standards of independence so that, when we confront a difficult case where intuitions are unsure, the analysis can guide our decisions of justification.

STICKS AND STONES

First, an easy example, an example in which there is little trouble in deciding whether the evidence used is good, credible evidence and the test of the theory gives good justification. Consider observation of a medium-sized, nearby, solid, opaque object such as a stone, a book, or a tree. These are the exemplars of what it is to be an observable object, and they are the most basic line of cold, hard evidence in science. What can be said of the independence of this evidence?

To evaluate an observation as scientific evidence we must be clear on the informational content of the claim. How it will function as evidence will depend on how the sensation is described. Pointing at a plainly visible stone, for example, and saying simply "I see it" or "Yes, there it is" is too ambiguous a claim to be useful. Does it mean the observer sees the atoms that make up the stone, or the whole stone itself, or perhaps the spirits that inhabit minerals and force them toward the ground, or even the essence of pain that splashes out when the stone hits you on the head? We have to know *what* it is that is observed if the claim can be evaluated for reliability and if it is to make a contribution as proof of other claims. The observation report must be of the form "I see that _____ is _____ ." In this case it might be "I see that it (this thing before me) is a stone" or, getting some action into the example, "The stone is falling with increasing speed."

To be used as evidence the observational claim must be accountable. It must be justifiable as reliable by an account of the viewing conditions and an understanding of the causal interaction between the object and the viewer. Not just any report from the senses is acceptable as evidence in science, only accountable reports. This does not mean that with each observation scientists must in fact present the justification. It means only that with each observation scientists *can* present the justification. The validation must be available, though not always displayed. We do not go around proving every claim we make, but we could if forced to. At least someone could. It's not that each individual observer must be able to validate his or her observational claims. People unaware of theories of optics can nonetheless render accountable, credible observations of stones. But the community of scientists must have a handle on the accountability of even these basic kinds of observation. It's not that the justification appeals to more basic, better-justified claims. The observation of the stone does not inherit its justification from more securely known theories of optics. The justification derives from the nature of the fit between claims, not from a privileged weighting of any one claim or any set of claims.

To answer the question of objectivity of the observational evidence about the falling stone we must see how the claim is used. It must be clear what the observation is evidence for, since only by knowing this can the feature of independence be evaluated. To further specify the example, then, consider the likely case in which the claim about the falling stone is used to prove a hypothesis about gravity, some claim perhaps about the comparable rates of free fall for a stone and a sponge. If this is the case, then the accounting theories, claims about perception, optics, the reflective properties of the stone, have no connection to the theory being tested, potentially claims about the force of gravity, and kinematics. In other words, it is a case of independent evidence, that is to say, objective evidence.

This is no surprise at all, and that is as it should be. There is no reason to offend our intuitions in such a clear case as this, and it is meant to be generally representative of cases of observation of midsized, nearby objects in which the report is of manifest information such as approximate size, color, shape, motion, or simply the presence of the object. This excludes observational claims, derived from the same sensations, of atomic composition or perhaps even the mass of the same object. To get this kind of information out of the observation one would have to put it in by oneself by way of the accounting theories. The point of this example is to show that those things we take to be obviously observable, sticks and stones, books, and the like, are within the class of things about which we can make entirely independent observation claims. Our intuitions about the clearly observable, in other words, match results of an evaluation of the feature of independence. This is not meant as a validation of intuition or to make you feel good about common sense. If either side is to gain credibility from the agreement, it is the analytic, independence side. The idea has been to show a match between intuition and analysis in cases where both intuition and the analysis of independence are easy to evaluate so that in harder cases, cases in which intuition is unsure, the analytic tool of independence can be used as a tool to evaluate objectivity and to extend our understanding.

FOSSILS

Here is a slightly harder case, observing fossils. These traces of organisms long dead form a crucial part of the evidence for the theory of evolution of life on earth. The evolutionary account makes definite claims about the forms of life at different times in the past, and insofar as some organisms leave clear images of their form preserved in sedimentary rock and easily seen by us, the theoretical claims have a seemingly straightforward source of observational evidence.

The shape of fossils is a manifest property. Anyone can have a look and report on the main features of form. Furthermore, appeals to mineralogical and chemical descriptions of the fossilization process will attest to the accuracy of the image. The process introduces only minimal distortion in the deposition and preservation of fossils. As we see them in fossils is pretty much as the organisms looked back then.

Back when? How is the information on the age of a fossil made available? The reliable dating of the fossils is an essential aspect of their value as evidence for the theory of evolution. At least a relative dating scale of fossils must be established so we know the order in which organisms were alive, since it is the progression of form that is the crucial claim of the theory of evolution. But, unlike coins, fossils do not have their dates stamped into them. There will have to be a theoretical accounting for the dates of fossils.

166

The most generally applicable method for determining the date of a fossil is to date the stratigraphy in which the fossil is embedded. Fossils are found in layered deposits of sedimentary rock. The layers, such as are visible in the Grand Canyon, are mineralogically as well as visually distinct. They have been deposited over time by oceanic silts and erosional runoff, each layer petrifying before the next is deposited. Thus, the depth of a layer is indicative of its age. The deeper it is the older it is, and so too with the fossils found in it. But how much older? The temporal order of the fossils, as representatives of past forms of life, must be quite long-term if it is to be consistent with the evolutionary mechanism of change. That is, the bottom layer and its fossils must be not just older than the top but *much* older. The bottom must be millions of years older, not just a few hundreds of years or a few days. Some measurement of dating is necessary if the fossils are to be evidence of evolution. Ordering is not enough.

The dating of strata as a means of dating the embedded fossils is a key link between the theory of evolution and the evidence in fossils. It is also a common point of dispute between advocates of the evolutionary account and opponents, principally those advocating a creationist story. The creationists charge circularity in the dating of fossils, a circularity that they claim impeaches the fossils' evidential worth. The fossils are dated by their location in the strata but, the creationist charges, the strata are dated from information of the fossils that are found in them. This would mean that there is no objective information on the dates of fossils. You put in what you want, the objection assumes, whatever serves the evidential interests of your theory.

This would be a serious objection to the fossil evidence for evolution if it were true. It is not true. There are in fact ways to determine the age of sedimentary strata without reference to fossils. Radiometric dating is the most definitive, and briefly, it works like this. (You want to understand how it works and to see its theoretical heritage so that you can evaluate its independence from the theory of evolution.) Some of the elements in the earth are radioactive. They decay; that is, they gradually and spontaneously turn into other stuff. They do this at known rates. For a radioactive element X that decays into another element Y, it is well known from physics how much time it will take for half of the deposit of X to turn into Y. This is the half-life of element X. Applied to geology, a sample of rock that includes some of element X can be measured for the relative amounts of X and Y. In a sealed deposit where the only source of Y is by decay from X, the ratio of amount of X to amount of Y is information as to how long X has been sitting around and decaying. It is the informational basis for an estimate of the age since formation of the rock.

This dating technique only works for igneous rock, the originally cooled and solidified deposits of molten magma. Sedimentary rock, the bed of fossils, is formed in natural processes of grinding up and eroding the igneous and metamorphic rocks. The pulverized minerals are redeposited and solidify into new rock. Radiometric dating does not reveal the age of a deposit of sedimentary rock because the minerals involved are reused, already solid and already partially decayed. But it is exactly the date of deposition that is needed since that is the date of the embedded fossils. Luckily, there is a way to use radiometric dating of igneous rocks to determine the dates of sedimentary deposits and thereby, in a noncircular way, determine the dates of the embedded fossils. Sedimentary strata are often found intermingled with igneous deposits. Sometimes there are igneous intrusions under or over layers of sediment, flows of magma from the sides, and so on. By dating the various bodies of igneous rocks and by determining the sequence of deposition of the various sedimentary and igneous rocks, one can bracket the dates of the sedimentary rocks. In the simplified picture shown in Figure 9.1, for example, the dates of igneous deposits A, B, and C can be established by radiometric techniques, and this information can be used to put limits on the

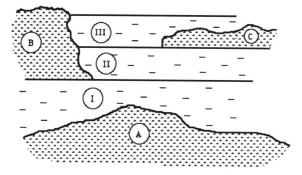

Figure 9.1. Sedimentary deposits (I, II, III) and igneous intrusions (A, B, C) in an orientation that allows dating the sedimentary deposits by using the dating of the igneous rocks.

dates of the sedimentary deposits. Layer I is younger than intrusion A but older than B, and so on.

The point is that the dates of the fossils found in sedimentary strata such as layers I, II, and III can be determined by radiometric analysis of the igneous intrusions. This is independent dating and independent evidence for the theory of evolution. It is using theories of physics (to describe the mechanism of radioactivity) to account for evidence used in behalf of biology. And there is nothing special in its being *physics* that accounts for the evidence. What is important is that it isn't evolutionary biology.

To complete the strategy for dating rocks and fossils we should note that not all sedimentary deposits are intruded by datable igneous rock. To date these isolated strata, geologists note characteristic features of strata they have dated elsewhere, features such as the types of fossils embedded, and look for these features in the isolated strata as indications of similar age. Thus fossils already dated by reference to radiometric data are used as reference to date otherwise undatable strata. This is perhaps the source of the creationist's confusion. But it is clearly not a case of circular reasoning or tight collusion between the dating of stratigraphy and the dating of fossils. There is a solid, independent basis in radiology for the information about dating.

The moral of this story is that, whereas the observational evidence of fossils is very heavily dependent on theory, the theories used are independent of the theory of evolution. Using fossils, including information on their ages, is not a circular, self-serving test of the theory. It is a case of independent, objective evidence.

CALORIC

The caloric fluid was a theorized entity invoked to explain and predict phenomena of heat. The presence and flow of caloric could account for the presence and flow of heat, as described in Chapter 1. An object that becomes hotter does so because of an increase in the contained amount of caloric and a resulting increase in caloric pressure. And just as water always tends to flow from a source at higher pressure to one at lower pressure, so does caloric tend to flow from hotter objects to colder. Caloric fluid, like water, seeks its own level. And like water, the flow of caloric can be used to do work, run machines, and generate electricity. The flow can of course be blocked. Thermal insulators are simply materials that are impervious to the caloric fluid. In sum, the caloric fluid can comfortably explain a lot of the manifest phenomena of heat.

The fluid itself, though, is never directly observed. It is imponderable; that is, it presents no tactile sensation, and it is invisible. Caloric fluid can be experienced only through the effects it has on things we can see and feel. Claims about caloric must be tested by their consequences, and in this sense the caloric fluid is like every other theoretical entity such as atoms, germs, or genes.

Many of these entities can be imaged, that is, indirectly observed with the aid of an appropriate machine. Science and its companion technology learn to observe more and more things with the use of various scopes. We have microscopes, telescopes, and oscilloscopes; is there such a thing as a caloriscope? Can the caloric fluid be observed in any indirect way, even if it involves lots of physical and theoretical steps? To answer this we should be specific as to just what information is putatively observed. Take, as a particular case, a claim to observe, if indirectly, that some caloric is flowing.

Here is how a caloric scientist might produce an image of the fluid in motion. Use as the imaging device, the scope, a manageably sized block of any ordinary stuff. A block of aluminum will do nicely. Now set it on the stove and turn on the burner. The block expands, visibly so, if you look closely and the original block is big enough. This process of expansion is an image of the process of caloric flowing into the block in the sense that the features of the expansion map exactly onto features of the flow. More flow causes more expansion. Stop the flow and expansion stops. Reverse the flow, expansion gives way to shrinking. You could even carefully connect one end of the block to a sensitive meter whose needle tips as the size of the block increases, and you have yourself a caloric flow meter. Or you could fix an ordinary thermometer into the aluminum block. Now the fluid mercury in the glass tube actually flows in harmony with the flowing caloric. This is a flowing image of flowing caloric. The mercury is not the same stuff as caloric, but then neither is the photograph of a DNA molecule as imaged by an electron microscope the same stuff as the molecule.

As with any other indirect observation, there is a series of interactions that map the features of the specimen onto features of the image. The reliability of the image is based on an understanding of these interactions, that is, on the accounting theories. In the case of the caloriscope the interaction is powered by the repulsive force felt between all particles of caloric. Whereas all massive objects attract, all caloric particles repel. So when the fluid inhabits an object the caloric particles spread out evenly to get as far away from each other as they can. They occupy the spaces between the aluminum atoms of our block and if more caloric is added the repulsive caloric particles must push for more space. This causes the expansion. In fact, if enough caloric is added, if, in other words, the object gets hot enough, the repulsive caloric particles will push hard enough to break the bonds of attraction between aluminum atoms, causing it to lose its rigidity. It melts. But this is getting beyond our experiment. What is important is that there is a clear interaction link from the caloric fluid to the expanding block.

How well does this observation play as evidence for the caloric theory? That theory predicts that caloric fluid will flow from a hot object, such as the stove, to a cold object, such as the aluminum block. But if this prediction is tested with our caloric flow meter it will not be a case of good evidence or objective testing. The reason is that the accounting theory, our guide to the accuracy of the meter, is practically nothing but caloric theory. Most importantly, the caloric theory plays an essential role in the accountability of the observation. We must put caloric theory into the account of how the image is formed in order to get caloric information from the observation. The story about repulsive particles, more entering the block, and the resulting expansion plants the information of flowing caloric. It is a clear case of a theory sponsoring its own evidence, not unlike a newspaper trying to validate its stories and build its credibility by printing the words "All of this is true" at the top of each page.

To summarize the story about this evidence for the caloric theory, the problem is not that the observation is very indirect. Lots of good observational evidence for lots of good theories is very indirect. Nor is the problem that the observation is theory-laden. All observation is theory-laden. It's not even that it is laden with a weak, unproven theory. That might make the evidence speculative and in want of more validation, but it would not impeach the evidence outright. The problem is that the observation is laden

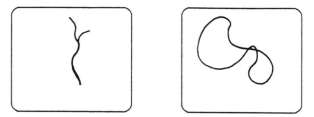

Figure 9.2. Electron microscope images of two DNA molecules, one that is split (*left*), another that is a closed loop (*right*).

with caloric theory. As evidence for caloric theory, then, it fails on evaluation of independence. It is not objective evidence.

ELECTRON MICROSCOPES

It is time to consider observations made with complicated machines. An electron microscope is built and declared reliable under the influence of theoretical claims about entities we never see, electrons. Is there any reason to think that such a complicated device, on the authority of claims about unobservables, can deliver credible, objective evidence? As always, it will depend on what one claims to observe, that is, on the informational content of the observational claim, and on what the information is used as evidence for. This is the moral of the story about evaluating evidence.

As a specific example, consider the case of using an electron microscope to observe a DNA molecule. This happens all the time. An electron microscope can image the global shape of the molecule and produce pictures like Figure 9.2.

Looking at the picture on the left one might claim to see that this particular DNA molecule is split, whereas the picture on the right shows a molecule that has a closed-loop structure. These sorts of observations might be used as evidence for claims about the biological effects of the shape of DNA molecules, claims of the

173

form that if the DNA molecule is split this will cause such and such manifest features in the offspring. The biologist can observe these features but cannot conclude from them that the DNA molecule is split. This would be relying on the link that is supposed to be demonstrated. A more direct demonstration of the link requires a view of its causal end, a view of the split DNA molecule itself. This is the need filled by the electron microscope.

To assess the objectivity of the DNA evidence we must know the accounting theories. We have to know of the theoretical account of how atoms in the molecule, as in any other material, interfere with the flight of electrons and how the electrons are scattered. (It may be a bit more complicated in that the atoms of the DNA molecule bond to heavier atoms of a stain that has been added and the atoms of the stain do the scattering.) Theories keep track of the subsequent behavior of the scattered electrons, describing the delay (called a phase shift) they experience in their encounter with the specimen, the alteration in their flight path, their deflection by magnets, and eventual reunion and comparison with those uninhibited electrons that were not scattered by the specimen. The information on scattering and where it took place is directed onto a photograph or a television screen. This is a complicated tale, but there is nothing in it about the biological function of the DNA molecule or the effects of a DNA molecule being split. It is, in other words, independent evidence. This makes it objective evidence.

DNA molecules are of course not the only things that can be imaged by an electron microscope, and it will be informative to consider another example. Single atoms have been imaged by electron microscopes. The picture on the screen is just a dot in a fuzzy background, not unlike the telescopic image of a lone, distant star. The observational claim could be simply that there is an atom present, or perhaps that there is a heavier atom in a crowd of lighter ones. Such an image and such a claim might be entered as evidence against atomic skeptics to prove that there really are such things as atoms.

This is the last time I will ask this question, but what are the accounting theories in this case? Since it is again a use of an electron microscope, the accounting theories are the same as in the DNA case. There must be an understanding of the specimen as an effective scattering target for electrons, an understanding that comes by way, as it did before, of atomic theory. That is, there will be an account of the atom with its dense, massive nucleus as the cause of the disturbance to the flight of electrons. And there will be the theories about the behavior of the scattered electrons and so on.

To say that the dot on the screen is not an artifact of the machine, we rely on the theoretical account of electrons. But to say more about what it *is*, to say what the dot represents, to say anything *atomic*, we must know already about the relevant features of atoms that cause the dot. This is more like a case of learning about the structure of DNA through its biological effects, the features of offspring, than the case of observing the DNA with the microscope. We have to use theoretical information on atoms as the cause of what we see in order to say that what we see is an atom. Atomic theory is a necessary part of the accounting theory in this case. This quashes the independence of the observation and makes it less objective as evidence.

THE BEST OF INTERNALISM

This way of assessing the credibility of evidence, of evaluating objectivity as independence in the account, is a way to deal responsibly with the predicament science is in. There is an accessibility gap between theory and evidence, that is, between the information of interest and the information that is available to us. The challenge is to use features of the information we have to indicate features of what we do not have. The best way to meet the challenge, as with a society that governs itself, is with a system of checks and balances. Science does not speak in a single voice, and we can exploit that diversity in the way we credit objectivity in the process of justification.

10

SCIENCE AND COMMON SENSE

Science is self-conscious common sense.
— W. V. O. Quine, *Word and Object*

NOW is the time to summarize. This final chapter will weave together the main ideas from the preceding pages and present the short, direct answer to the question of scientific method and the justification of theories. More like a travel brochure than an encyclopedia, the point here is to sketch the big picture of the scientific process and to locate its source of believability. And like a travel brochure this chapter will not give you anything new, only old news neatly packaged.

Now is also the time to draw the parallel between scientific reasoning and our everyday ways of coming to know and understand the basic features of the world around us. The predicament, the accomplishments, and the standards by which science achieves knowledge of the underlying mechanisms of nature are not so very different than the predicament, accomplishments, and standards that you and I deal with in simply knowing about objects in the living room, the weather outside, or when to cross a busy street.

Reasoning in science is like reasoning in general. In the context of science, the processes of discovery and justification are slowed down and hence easier to follow in detail. A single step in justifying a scientific theory may take days, or weeks, or even years to carry out. There are no snap decisions like the ones we are forced to make in life on the street, deciding when it is safe to cross and when it isn't. The scientific process may run more slowly, but its parts and its movements are essentially similar to those of common sense.

Justification in science is also a more openly public process than is justification on the streets. The standards of justification are more explicit. It is more frequent in science than it is elsewhere in life actually to produce one's justification rather than settling in quiet confidence that the case could be made. The more explicit process makes it so that we can more easily see the steps of justification and more clearly visualize the vital functioning parts. Science, after all, is a very communal activity where ideas and standards are shared and compared among people. This can only be done if the ideas are articulated. In contrast, many of the beliefs we hold as individuals are the products of tacit thought. Much of the justifying of common sense, the comparison among beliefs to see that they make sense, is tacit. Doing science forces the process out in the open. It is, in a sense, visible, articulated thinking.

Metaphorically, the scientific process is like a magnified image of our everyday processes of thinking and knowing. The scientific process is more open and accountable than are our private thoughts. It is also more careful to obey its own standards of good reason. Now that we understand what these standards are we can apply them to life on the street and our more mundane struggle for knowledge of the external world.

SCIENCE

Here is a retelling of the story, a quick but informed sketch of the scientific image. It will highlight those features of the process that are the source of respect and credibility of science, those standards that must be carefully met if we are to do things scientifically.

The standards of science require a respect for the availability of information used to formulate or justify descriptions of the world.

178

There can be no appeal to the unknowable. Justification cannot rely on factors beyond our understanding, lest the description descend to mysticism. A phenomenon explained as being just magic or supernatural is not explained at all. The suggestion is that its causes and control are simply beyond our comprehension, and this sort of appeal to unknowable, inaccessible information is the antithesis of science. Organizing events under the ubiquitous will of a god, to cite another approach to describing the world that disregards the scientific standard of using only available information, puts the account beyond our comprehension and even off limits to our attempts to understand. It is mysticism to describe and explain the world in terms of the unseen, the unseeable, and the unknowable. It is certainly not what we expect of science.

The other side of this same coin, this respect for the availability of the information used, is that science can have no appeal to or reliance on the brute facts of the world. This too would be a kind of mysticism since it asks us to describe or justify theories in terms of things as they are in the world independent of any descriptive, conceptual, theoretical system. To avoid this and to respect the standard of accessibility of information, science must rely on information from within the theoretical system. This is what is meant by saying that factors of justification must be internal.

We have seen two reasons why science is constrained to this internal perspective, two reasons why the justification of theories is dependent on a theoretical system. The first was underdetermination. Explanation and confirmation, as activities of justification of theories, must rely on internal features and relations to other theories to be decisive. The observational evidence alone is never sufficiently informative to settle questions of *best* explanation, confirmation, or refutation. This shows the unavoidable network structure of the theoretical system and its justification. Theories confront the evidence only under the influence of other theories. Even if the observational claims were pure and themselves beyond the influence of theory, they could only be explained or used as confirmation in the context of a network of a theoretical system.

But of course the observational claims are not pure in this sense, and this is the second reason that science is forced to confront the world from an internal stance. Each observation bears a theoretical component and is itself a part of the theoretical system. There is no piece of evidence that is not theoretical. As soon as we step outside the reaches of the theoretical system we are among the useless and uninformative. Brute sensations are not evidence and do not contribute to science. Their informational content is accessible only when linked to the system.

Thus the same respect for the accessibility of information that prevents mysticism from being passed as science also prevents a reliance on foundational, external facts to enter any enterprise we regard as scientific. Making sure that the topic of conversation is within the bounds of available information guards against a variety of sins.

In addition to this standard of availability of information, science functions under a standard of objectivity as openness. This is a feature of the process of science rather than of the claims science makes. It is descriptive of how things are done, in other words, rather than of the results of the process. The testing and acceptance of theories are carried out with an openness to potential refutation. It is unacceptable to simply *ignore* observations that seem to refute particular theories or to ignore contradictions among theories in the system. These things must be accommodated by adjustments to the network if only by explicitly discounting the conditions under which the observation took place or by changing aspects of the accounting theory that grants the observation its relevance to theory. Facing up to the evidence by making even these adjustments will have ramifications within the theoretical system that will be constrained by coherence. You

180

cannot deny an accounting theory or a claim about conditions unless the denial is consistent with the remaining, or perhaps readjusted, claims in the system. Being open to evidence, as this standard of objectivity requires, forces adjustments to the system, and coherence insists that not any old adjustments will do. In this sense the ongoing justification process is forced out of our personal control. It is kept more objective. The point is that a *scientific* process of justification cannot be insular. It cannot keep a theory free of refutation by hiding it from refutation, whether by refusal to check other claims in the system, observations or theories, or by refusal to acknowledge those that conflict. The coherence of the system must be maintained in a dynamic environment of challenge.

There is also the feature of objectivity as independence. This speaks for the quality of the evidence used in justification. It is not just seeking and using any old evidence that counts as objectivity in science. It is better to seek independent evidence, evidence whose accounting is uninfluenced by the theory it is being used to test. This aspect of objectivity enhances the credibility of the case.

These are, then, the essential features of the scientific process. It is a process governed by the restriction of accessibility of information and by a standard of objectivity as openness to challenge and independence of evidence. With these essential constituents, the dynamic process of scientific justification can best be described as an ongoing negotiation between theory and observation.

Here is how it works. Theories must be informed by observations. This is no surprise. Science relies on observations of what's happening on the manifest surface of the world to suggest the content of theories about what's going on beneath the surface. Observations are the supply of particular claims from which to draw inductive generalizations. Observations present the phenomena of confirmation that shape the theoretical system under the demand of consistency and explanatory relevance. Observation guides theorizing.

Observations must also be informed by theories. This too should be no surprise, at least at this point. Theories are the source of accountability of observations by drawing out their informational content in a full, descriptive presentation that can function as evidence and make relevant contact with theory, and by structuring the guidelines of reliability of reports from sensation. Theory guides observing.

This reciprocal dependency calls for a method of negotiation and give-and-take between theory and observation. In fact, it is a method of mutual influence and negotiated agreement among all the claims in the theoretical system. It is not so important to distinguish the observational from the theoretical as it is to accommodate all claims into a consistent, well-connected fit. This is coherence and it is the principal global constraint that structures the justification of scientific claims. Added to the basic aim of coherence is the standard of independence of evidence. Negotiation is more challenging and agreement is more significant when they involve parties with different aims. In science, objectivity is served by insisting that the theoretical influence on observation be by theories whose aims will not be served by the outcome of the particular observation.

It is not that a theoretical system that is kept under the guidelines of coherence plus independence will show nothing but true claims. The standards cannot be made that strict. But as the process goes on, as the system is revised under the pressure of coherence and independence, revised to accommodate new observations and to settle old inconsistencies, the system will be nudged in the direction of truth.

These claims about the predicament and process of science are conceptual claims about the enterprise of acquiring knowledge about the world. They are not merely a report on how it happens to be done. This view of science is based on the necessary limitations of information, the nature of observation, and the role of independence. No part of the account has been drawn from the features of any particular science or any particular kind of entity science might study. This is what allows for the generalization across the natural sciences. It is not a specialized account of how to theorize and justify claims specifically about evolution, or about gravity or light. It is more generally an account of theorizing and justifying claims about what is not manifest in our encounter with the environment, and it has been guided by the conceptual constraints on expansive knowledge in general. This has allowed for the variety of examples drawn from physics, geology, astronomy, biology, and archaeology.

Such a general account of knowledge beyond the apparent should apply as well to more humble, everyday cases of our coming to know things about our environment that are not immediately manifest.

COMMON SENSE

What we take to be responsible, justified knowledge in our day-to-day confrontations with the external world is similar to what has been described above as responsible, justified knowledge in science. There is certainly a similarity in both the aims and the informational predicament of the two enterprises. In both cases the aim is to know more than is immediately apparent in the evidence. Expanding our knowledge beyond our impoverished supply of data is the business of science; it is also the nature of knowledge day to day. As individuals in the world we learn of our environment through our perceptions. What we perceive, of course, are episodic images of objects that change shape and color as our own perspective and viewing conditions drift, images that flash in and out of view as our attention and location change. Insofar as we want to know about the world itself, about the trees, our friends, and oncoming traffic, we want information that goes beyond our perspective-dependent, fleeting view. We want to know about the enduring objects themselves. We want, in other words, to know about more than our own perceptions, but the perceptions present the available information. The task of knowledge is to use the perceptions, the apparent information, as evidence for claims about the objects. The task is to expand our knowledge beyond the evidence at hand. We are not satisfied with claims about the episodic, condition-dependent, subjective world of perceptions. We want to know, that is, we want to make *justified* claims and not wild guesses or capricious opinions, about enduring objects. And of course this more ambitious knowledge must be informed by the more humble evidence of perceptions. In sum, the aims and predicament of an individual's knowledge of the external world are like those of the enterprise of science. We are ambitious knowers.

It should be of little surprise that this similarity of situation brings a similarity of method and of standards for science and common sense. In our daily confrontations with the world, as in science, justification by correspondence is not an option. The facts of the matter, the objects as they are and not as they are perceived, to which our beliefs are supposed to correspond, are not available. This is our predicament.

We are stuck with coherence. The method we must use to justify our beliefs and claims about the external world is one that pursues coherence in a dynamic process of open exposure to the perceptual evidence. "Let's have a look," says common sense, when there is some doubt about what is going on in the world. That is, let's consider the evidence and be open to possible refutation of our beliefs. Let's make sure that our beliefs about objects in the world can accommodate our observational beliefs that we get through perception. Let's check that observational claims are consistent with and even explained by theoretical claims.

184

But no conscientious investigator of nature, even in the mundane knowledge of the normal environment, is a slave to observation. In life as in science we have to be on our toes to be sure that what we seem to observe, what we claim to observe, is an accurate report on what is really happening in the world. As individuals we must be willing to doubt and discount what we see if it too seriously offends our system of beliefs about how things work in the world. We are too smart to be taken in by magic tricks, for example, because we know that this kind of stuff just doesn't happen in the world. People don't disappear in a puff of smoke, and so a solid, well-entrenched network of beliefs directs the justification process to devalue the credibility of these sights. These basic beliefs about the world, our commonsense theories, influence the observations we make as individuals. Our observations are theory-dependent just as our theories are observation-dependent.

The standard of acceptability, then, of observational and more theoretical claims is coherence between and among both. Beliefs about the world gain acceptance by their consistency with and explanation of perceptual claims, and perceptions are believable in part because they fit without much stretching into the network of beliefs. In addition to this we impose a standard of independence to seek out some claims as more significant than others for comparison. The significance is not a product of some particularly foundational status of these beliefs as being directly observational or antecedently better confirmed. The significance is a contextual feature of independence. We give higher credence to evidence that is accountable by means that are independent of the particular claim it is evidence for. This is a guide to the status of evidence in a court of law, evidence of our perceptual systems, and evidence in general. It is less likely that claims are being fabricated or that we are hallucinating if disparate, independent accounts intersect on the evidence and agree. In science as on the streets, objectivity as independence further guides our justification of what we claim to know about the world and gives us a better chance at accepting what is true and rejecting what is false. Between science and common sense are only shades of difference.

RISKY BUSINESS

Science, like life on the streets, is risky business. Any time expansive claims are made about what is going on beyond the manifest there is the risk of being wrong. It is the nature of ambitious knowledge that there can be no certainty. Anything interesting, anything that is more than just a description of the obvious, will be more or less uncertain. The important task of the scientific process is to move our theoretical understanding in the direction of being less uncertain. The accomplishment of the scientific method is not to eliminate the risk but to reduce it, to choose theories more likely to be true. Within the constraints of available information you do the best you can. No theory is dead certain, but some are better justified than others, and those are the ones we want. To help get what we want we must continue to articulate and apply the standards of justification.

No one has finished reading the book of nature, and no one ever will. This is because there is an endless supply of observations yet to be done, lines in the text yet to be read. At any stage of the study there will be a multitude of interpretations that can be fit to the text so far encountered. But there are ways to decide, among these multiple readings, which of them are more likely to be true. Apply the constraints of coherence and independence, and of course, keep reading.

186

GLOSSARY OF TERMS

Accounting theory: A theoretical claim used to attest to the reliability of an observation or to describe the informational content of an observational claim is being used as an accounting theory. It is a designation of how a theory is used rather than of any intrinsic features of content or scope of a theory. As an example, theoretical claims about the behavior of electrons are used as accounting theories in the case of observing a DNA molecule (or anything else) with an electron microscope.

Auxiliary theory (sometimes called an "auxiliary hypothesis"): The background knowledge required to draw testable consequences from a hypothesis is found in the auxiliary theories. No hypothesis alone entails testable predictions, so auxiliary theories always participate in the testing of a hypothesis. For example, the hypothesis that the solar interior is a furnace of nuclear fusion has the testable consequence that solar neutrinos are showering the earth, but the consequence can be drawn out only with the theoretical knowledge that fusion produces neutrinos. The claim about the production of neutrinos is an auxiliary theory in this case of testing. Like accounting theories, the designation of "auxiliary" is descriptive of how a theory is used in a particular case but not of a theory's content or justification status.

Coherence: A system of beliefs (or claims, or theories) is coherent to the extent that the individual beliefs are related to each other by implication and explanatory relevance. Coherence is a global feature of a belief system. It is composed and evaluated in terms of relations between beliefs, unlike correspondence, which is a relation between a belief and an object or event in the external world. Coherence presupposes consistency. As an example, the set of claims, "The moon is full," "All dogs are vicious," "Smoking is an unhealthy activity," is a consistent set (since no claim contradicts any other), but it is not as coherent as the set, "The moon is full," "The moon affects the tides," "The tide is higher than usual." This latter system of claims is consistent, and it has a high degree of coherence.

Correspondence: A belief (a claim, a theory) corresponds to a fact if the belief accurately describes the state of affairs in the world. Evaluating correspondence would require an awareness of and comparison between both the belief and the state of affairs in the world. Our inability to gain awareness of the state of affairs except as it is represented through the belief is what makes correspondence between beliefs and the world impossible to evaluate. Correspondence may be what we mean by "the truth," but it cannot be the way we evaluate the truth.

Covering-law model: This is a very general, formal model of explanation in which an event is explained if it is shown to be an instance of a general law of nature. The law, in other words, covers the particular case. There are several varieties of covering-law models with various restrictions on the nature of the explanatory laws and of the way they cover. The deductive-nomological (D-N) account, for example, requires that the event to be explained be a deductive consequence of the law.

Deductive: One or more claims (the premises) deductively imply another (the conclusion) if the truth of the former guarantees with certainty the truth of the latter. The inference from "This dog is not vicious" to "Not all dogs are vicious" is a deductive inference because if the first claim is true then the second claim *must* be true as well. The term "deductive" describes the certainty of the link between claims. It has nothing to do with the truth of a single claim or with the generality or singularity of claims or arguments.

Deductive-nomological model (D-N): This is a particular, and particularly common, form of covering-law explanation in which the law that covers the event (plus claims about the particular conditions) deductively entails the statement about the event being explained.

Empiricism: This is a kind of antirealism that claims that scientific theories are in fact true or false in what they say about both observables and unobservables, but we can have justified knowledge only about what they say about observables. The responsible position to take with regard to claims about what cannot be observed is a position of agnosticism: Withhold belief in those claims that are beyond the limits of knowledge. Note that the term "empiricism" is used more generally by some authors to cover all antirealist positions, both suspended belief about unobservables and belief that they do not exist, that is, any position that bases justification firmly on observation.

External: This is short for "external of our descriptive, theoretical system." External features of a theory would indicate how the theory relates to things that are not part of or influenced by the general network of theories and descriptive claims about the world. Correspondence, for example, is a relation between beliefs and external facts. "External" does *not* mean outside of a person's body or a person's mind or beyond an individual's intellectual grasp. It means outside of and independent of the system of theories held collectively by people.

Hermeneutic: As used here, this is a term that describes a method of acquiring and justifying knowledge that goes beyond the evidence. It is derived from the particular method of interpreting texts written in an unfamiliar language. The distinctive feature of the hermeneutic method is the recognition that the parts of an interpretation (individual passages) have meaning only from their context in the whole text, and the whole interpretation is given meaning by composition of the parts. Thus understanding of the whole guides the understanding of the parts, and the whole is understood from the parts. This is the hermeneutic circle. The method proceeds by hypothesis of meaning being tested for consistency and coherence of the interpretation it generates.

Hypothesis: The particular theory that is being tested in a certain case is the hypothesis in that case. This term is used to identify a theory (or theories) that is in need of testing. Thus a hypothetical claim is made without much commitment to its truth. It is a suggestion in want of confirmation.

Hypothetico-deductive confirmation (H-D): This is a very general, formal model of confirmation in which a hypothesis is tested by its deductive consequences. It is by no means a deductive proof that the hypothesis is true. Rather, it is a way of providing inductive evidence that the hypothesis is likely to be true. For example, the hypothesis that all dogs are vicious has the deductive consequence that the collie next door is vicious. I test this by testing the disposition of the collie next door. If she is vicious, then this is *some* small indication that the hypothesis is true.

Inductive: One or more claims (the premises) inductively imply another (the conclusion) if the truth of the former makes it more or less likely that the latter is true. Unlike deductive inference, inductive inference comes in degrees of strength. The inference from "No dog that I have met is vicious" to "No dogs at all are vicious" is an inductive inference because the truth of the first claim makes it likely that the second claim is true but it does not guarantee it. Any argument that leaves even the smallest chance that the conclusion could be false even if the premises are true is an inductive argument. The term "inductive" has nothing to do with the truth of a single claim or with the generality or singularity of claims or arguments.

Instrumentalism: This brand of antirealism claims that scientific theories are not intended to be evaluated as being true or false. They are only instruments we use to guide our thoughts and beliefs about the world we can observe.

Internal: This is short for "internal to our descriptive, theoretical system," and it is in direct contrast to the term "external." Internal features of a theory indicate how the theory relates to other theories and beliefs. Any bit of information that can be evaluated only with reference to theories or beliefs is internal.

Law: A claim that associates a general kind of thing with a particular behavior or property is a law. Laws identify natural kinds in the world and imply a causal connection between being that kind of thing and having the associated property or behavior. Claims about accidental associations and the circumstantial properties of such groups are not laws, even though they are generalizations.

Observation: Insofar as this activity is to contribute to science it should be more fully put as "observational claim" or "observational belief." An element in the descriptive system is observational to the extent that it is caused by a particular sensation. Observations are not the product of reflection. They occur to us spontaneously rather than through an act of comparing and drawing inferences from other ideas. Observations are beliefs we get as a result of direct impact by the environment, though they gain their importance as evidence to function in the process of science only with reference to theory.

Realism: One is a realist in the issue of justification of scientific theories if one believes that not only are theories definitively true or false in what they say about observables but, further, that we can in some cases tell when a theory is true even in its claims about unobservables. In other words, science can and does make justified knowledge claims about aspects of the world that cannot be observed.

Theory: Any claim about the external world, general or specific, well justified or speculative, is a theory. Theories are the elements of expansive knowledge in that they always make claims to more than is told by the evidence.

SUGGESTED READING

In General

Quine, W. V., and Ullian, J. S. *The Web of Belief.* New York: Random House, 1978. This is a very readable and provocative book about the holistic network of our beliefs about the world. It is more about knowledge in general than about science in particular.

Hempel, C. *Philosophy of Natural Science.* Englewood Cliffs, NJ: Prentice-Hall, 1966. This classic introductory text is the place to find clear, straightforward descriptions of the traditional models of science such as the covering-law model of explanation and the hypothetico-deductive model of confirmation.

Hacking, I. *Representing and Intervening.* Cambridge: Cambridge University Press, 1983. Slightly more difficult than Hempel, this is a more modern introduction to topics in the philosophy of natural science with a particular emphasis on the issue of realism and the role of experiment in science.

Boyd, R., Gasper, P., and Trout, J. (eds.). *The Philosophy of Science.* Cambridge, MA: MIT Press, 1991. This is a rich anthology of important papers on the main issues in philosophy of science. The entries vary in difficulty, but all show progress toward understanding science.

Kourany, J. (ed.). *Scientific Knowledge.* Belmont, CA: Wadsworth, 1987. This anthology of recent articles presents some of the most influential work of this century on such topics as explanation, confirmation, and realism.

Bonjour, L. *The Structure of Empirical Knowledge.* Cambridge, MA: Harvard University Press, 1985. This is significantly more difficult than the other books in this section. It is written by a philosopher, directed to readers with some background in philosophy, but it is the most complete and careful presentation that I know of the idea of coherence as justification.

192

Suggested Reading

Chapter 1. Theories

Carnap, R. *An Introduction to the Philosophy of Science.* New York: Basic, 1966. Chapter 23 of this readable book by one of the century's leading philosophers of science describes in plain terms the nature of theories in contrast to observations.

Chapter 2. Internal and External Virtues

Quine, W. V., and Ullian, J. S. *The Web of Belief,* chapter 6. Here is an explicit discussion of "virtues" of hypotheses, virtues such as conservatism, simplicity, and generality.

Hempel, C. *Philosophy of Natural Science,* chapter 4. Hempel discusses internal virtues of evidence and of hypotheses under the heading of "criteria of confirmation."

Chapter 3. Explanation

Hempel, C. *Philosophy of Natural Science,* chapter 5. This is as clear as any presentation of the basic form of the covering-law model of explanation.

Kourany, J. (ed.). *Scientific Knowledge,* part 2. This is a collection of contemporary articles on the nature of explanation in science, representing a variety of approaches to the issue, including the covering-law model and its variations.

Cartwright, N. *How the Laws of Physics Lie.* Oxford: Oxford University Press, 1983. This is somewhat technical, but essay 5 presents a good case for the specifically causal nature of explanation.

Salmon, W. *Four Decades of Scientific Explanation.* Minneapolis: University of Minnesota Press, 1989. Salmon is one of the most influential and most active writers on scientific explanation. This is a comprehensive and authoritative presentation of recent work on explanation, surveying a variety of approaches and conclusions.

Kitcher, P. "Explanatory Unification and the Causal Structure of the World." In P. Kitcher and W. Salmon (eds.), *Scientific Explanation.* Minneapolis: University of Minnesota Press, 1989. Kitcher makes a clear and persuasive case that a scientific explanation must demonstrate unification. The laws used to explain phenomena must be fruitful in the sense of being able to cover a variety of phenomena with a minimum of theoretical detail.

Suggested Reading

Chapter 4. Confirmation

Duhem, P. *The Aim and Structure of Physical Theory.* New York: Atheneum, 1962 (French ed., 1914). This is a classic source for a description of the necessity of auxiliary theories in confirmation and the consequences that crucial experiments are not possible and in general that evidence underdetermines theory.

Popper, K. *The Logic of Scientific Discovery.* New York: Basic, 1959 (German ed., 1935). Karl Popper is well known for the view that the essential feature of a scientific claim is that it is falsifiable, and this book is the manifesto of the falsifiability movement.

Glymour, C. *Theory and Evidence.* Princeton, NJ: Princeton University Press, 1980. This is a novel account of confirmation with a kind of internalism, called bootstrapping, in which some parts of a theory are used in the confirmation of other parts.

Kourany, J. (ed.). *Scientific Knowledge,* part 3. This section of the Kourany anthology contains a variety of articles on issues of testing scientific theories. There are short but informative works by Carnap, Popper, Duhem, and others.

Brush, B. "Prediction and Theory Evaluation: The Case of Light Bending." *Science* **246** (1989), pp. 1124–1128. This is a succinct account of the observations of the bending of starlight and their relevance as evidence to confirm the general theory of relativity.

Bahcall, J. "The Solar-Neutrino Problem." *Scientific American* (May 1990), pp. 54–61. The current troubles with finding the solar neutrinos and the considered excuses for why they have not yet been found are clearly told and illustrated.

Chapter 5. Underdetermination

Hacking, I. *Representing and Intervening.* This sustained and lively treatment of realism has lots of nice examples. Hacking distinguishes between realism about theories (we can know if they are true or not) and realism about entities (we can know if they exist or not) and defends the latter.

Leplin, J. (ed.). *Scientific Realism.* Berkeley: University of California Press, 1984. This is presently the best collection of recent articles on scientific realism. It is not always easy reading, but it is clearly the place to see the various approaches to answering the question of realism.

Suggested Reading

Nye, M. *Molecular Reality*. London: Macdonald, 1972. As a historical account of the various ways that Jean Perrin measured Avogadro's number, this is the story of the turning point in the scientific community's belief in the existence of molecules.

van Fraassen, B. *The Scientific Image*. Oxford: Clarendon Press, 1980. A very influential presentation of an empiricist view of science, this also contains a novel account of explanation. It is not an easy read, though.

Chapter 6. Observation

Churchland, P. *Scientific Realism and the Plasticity of Mind*. Cambridge: Cambridge University Press, 1979. Chapter 2 of this book makes the case that sensation is not evidence and that observation can contribute to science only if it has been influenced by theory.

Hanson, N. *Patterns of Discovery*. Cambridge: Cambridge University Press, 1958. This is where the phrase "theory-laden observation" came from. There are many enjoyable examples to show that even what we perceive is influenced by the theories we hold.

Scheffler, I. *Science and Subjectivity*. 2nd ed. Indianapolis: Hackett, 1982. In chapter 2 Scheffler acknowledges that what we observe is dependent on theories to set a system of categories in which to organize sensations, but he maintains that which category a particular sensation fits into will be determined by nature, not by the subjective influences of the observer.

Hesse, M. *The Structure of Scientific Inference*. Berkeley: University of California Press, 1974. The first chapter, "Theory and Observation," presents an account of observation claims, their need of theoretical support, and their revisability. The account is sensitive to many detailed concerns of philosophers, concerns, for example, about the meanings of terms and the epistemic status of universals.

Chapter 7. Blurring the Internal–External Distinction

Newell, R. *Objectivity, Empiricism, and Truth*. London: Routledge and Kegan Paul, 1986. All our beliefs and claims about the world bear an unavoidable internal influence, but there is an internal kind of objectivity that is an openness to evidence for or against one's beliefs.

Suggested Reading

Kuhn, T. *The Structure of Scientific Revolutions.* Chicago: University of Chicago Press, 1962 (2nd ed., with postscript, 1970). Arguably the most influential philosophy of science book of our time, this is provocative, easy to read, and easy to misunderstand. It introduces the notion of a paradigm as, among other things, an ensemble of concepts, presuppositions, and practices that inexorably influence our theorizing and observing.

Chapter 8. Coherence and Truth

Bonjour, L. *The Structure of Empirical Knowledge.* Difficult, but this has motivated much of what I have to say about the link between coherence in the theoretical system and the truth of the theories.

Holton, G. *Thematic Origins of Scientific Thought: Kepler to Einstein.* Cambridge, MA: Harvard University Press, 1973. Section 2 describes the development of the special theory of relativity.

Chapter 9. Objective Evidence

Hacking, I. *Representing and Intervening.* With its emphasis on the role of experiment, this shows a variety of examples of observations and their accounting theories.

Kitcher, P. *Abusing Science.* Cambridge, MA: MIT Press, 1982. Chapter 3 in particular is a clear account of the dating of fossils and their use as evidence for the theory of evolution.

INDEX

Index